SpringerBriefs in Environmental Science

SpringerBriefs in Environmental Science present concise summaries of cutting-edge research and practical applications across a wide spectrum of environmental fields, with fast turnaround time to publication. Featuring compact volumes of 50 to 125 pages, the series covers a range of content from professional to academic. Monographs of new material are considered for the SpringerBriefs in Environmental Science series.

Typical topics might include: a timely report of state-of-the-art analytical techniques, a bridge between new research results, as published in journal articles and a contextual literature review, a snapshot of a hot or emerging topic, an in-depth case study or technical example, a presentation of core concepts that students must understand in order to make independent contributions, best practices or protocols to be followed, a series of short case studies/debates highlighting a specific angle.

SpringerBriefs in Environmental Science allow authors to present their ideas and readers to absorb them with minimal time investment. Both solicited and unsolicited manuscripts are considered for publication.

Javid Ahmad Parray • Mohammad Yaseen Mir
A. K. Haghi

Enzymes in Environmental Management

 Springer

Javid Ahmad Parray
Environmental Science
HKM Government
Degree College Eidgah
Srinagar, Jammu and Kashmir, India

Mohammad Yaseen Mir
Centre of R & D
University of Kashmir
Srinagar, Jammu and Kashmir, India

A. K. Haghi
Department of Chemistry
Institute of Molecular Sciences
University of Coimbra
Coimbra, Portugal

ISSN 2191-5547   ISSN 2191-5555   (electronic)
SpringerBriefs in Environmental Science
ISBN 978-3-031-74873-8   ISBN 978-3-031-74874-5   (eBook)
https://doi.org/10.1007/978-3-031-74874-5

This Springer imprint is published by the registered company Springer Nature Switzerland AG
The registered company address is: Gewerbestrasse 11, 6330 Cham, Switzerland

If disposing of this product, please recycle the paper.

Contents

Chapter 1
Various Enzymes to Treat Resistant Pollutants in Wastewater: A Sustainable Practice for Environment

Abstract Degrading resistant contaminants in wastewater, such as oil, grease, pesticides, medicines, and plastics, is challenging for conventional activated sludge processes. Due to their widespread presence, toxicity, resistance to natural biodegradation, and other detrimental effects, these pollutants pose a significant hazard to aquatic habitats and organisms. Enzymes are suggested as high-efficiency biocatalysts for the removal of these harsh contaminants. The functions and uses of enzymes in wastewater treatment were the main topics of this review. It goes over the importance of different kinds of enzymes, where they come from, how to remediate resistant pollutants using enzymatic processes, identify and test the ecotoxicity of enzymatic transformation products, and how commonly used enzymatic wastewater treatment systems are. There are suggestions for future study areas and perspectives on the main difficulties with enzyme-based wastewater treatment.

Keywords Enzyme-driven wastewater treatment · Resistant pollutants · Immobilized enzyme systems · Enzyme-mediator

1.1 Introduction

Alarmingly high levels of various pollutants have been released into aquatic habitats in recent years. Traditional activated sludge, the most popular wastewater treatment technique, can eliminate most pollutants; however, removing resistant contaminants such as plastics, oils, greases, medications, and pesticides is more challenging. These pollutants mainly consist of organic micropollutants, oil, and grease (Fig. 1.1). Dairies, oil mills, slaughterhouses, and food waste are the typical sources of wastewater that contains oil and grease. The transfer rate of substrates, products, and oxygen will be hampered by oil and grease that floats on water surfaces. Grease and floating oil may result in a filamentous microbe bloom and poor activated sludge performance, producing floating sludge, poor sedimentation, and decreased sludge biomass. According to Sauvé and Desrosiers (2014) and Teodosiu et al. (2018), emerging pollutants or emerging concern contaminants include other pollutants that

© The Author(s), under exclusive license to Springer Nature Switzerland AG 2024
J. A. Parray et al., *Enzymes in Environmental Management*, SpringerBriefs in Environmental Science, https://doi.org/10.1007/978-3-031-74874-5_1

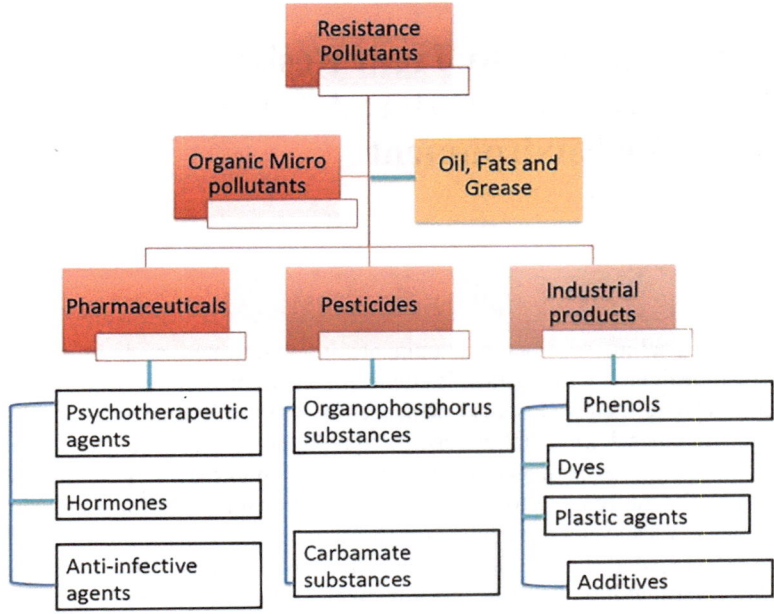

Fig. 1.1 Typical resistant pollutants in wastewater and their classification

are also referred to as micropollutants, such as plastics, medicines, insecticides, and personal care items. Because of their nanogram-to-microgram-per-liter quantities in the environment, these substances are difficult to detect or manage in most settings. They pose a severe hazard to the environment and all living things since they are usually poisonous, readily bioaccumulate, and resistant to natural biodegradation (Sauvé and Desrosiers 2014).

The two primary types of wastewater treatment techniques used nowadays are biological and physiochemical. Wastewater treatment uses physiochemical processes such as chemical oxidation, distillation, membrane-based separation techniques, and adsorption (Alshabib and Onaizi 2019). Regarding treatment, these approaches are highly costly and potentially exacerbate pollution and harm. Most contaminants in wastewater can be eliminated by biological processes, which are also more environmentally friendly. These techniques treated wastewater using bacteria, plants, and enzymes (Alshabib and Onaizi 2019). In the process, the organisms used in these technologies may break down or absorb pollutants. On the other hand, enzymes can function quickly and selectively, whereas plants and microorganisms are susceptible to some harmful contaminants from wastewater.

1.2   Enzyme Treatment in Different Wastewater Pollutants

1.2.1   Selection of Enzymes

The types and sources of enzymes significantly impact the efficiency of enzymes in wastewater treatment. Enzymes are particular, meaning they can only catalyze a specific process. For example, lipases can catalyze the breakdown of fat, grease, and oil, whereas proteases may break down proteins. Carbaxylesterases catalyzed the hydrolysis of ester bonds of carbamates, organophosphates, and other chlorinated organic compounds (Cummins et al. 2007). While haloalkane dehalogenases ought to be permitted to split the carbon–halogen bond of halogenated aliphatic compounds, phosphotriesterases can hydrolyze various organophosphates (Rezvani Romeh and Hendawi 2014).

Laccase and peroxidase exhibit excellent selectivity even if they contain various chemicals. According to related studies, horseradish peroxidase and Pleurotus ostreatus laccase can efficiently and selectively eliminate a few medications. While the Pleurotus ostreatus laccase required 20 min to transform diclofenac (DCF) completely and sotalol (STL), the horseradish peroxidase could convert APAP in about 4 h. Nevertheless, sulfamethoxazole (SMX), carbamazepine (CBZ), ibuprofen (IBP), and naproxen (NAP) did not alter either enzyme. According to a different study, enzyme sources significantly impact how well wastewater is treated. In their research, Daâssi et al. (2016) employed distinct fungal laccases to biotransform BPA, with the Coriolopsis gallica laccase exhibiting a higher oxidation rate than the other fungus laccases. According to Garg et al. (2020), mesquite peroxidase can remove over 90% of phenol from wastewater in 30 min. In contrast, peroxidases from potatoes, soybeans, and horseradish (Rajesh et al. 2017) require longer residence times to achieve the same degree of elimination (Al-Ansari et al. 2010). When 0.5 mL of crude soybean peroxidase was used for azo dye methyl orange treatment, it was able to achieve a maximum decolorization of 81.4% after 1 h of incubation at 30 °C with 2 mM hydrogen peroxide and 30 mg/L of methyl orange at pH 5.0. In the meantime, 1.5 mL of luffa peroxidase showed a maximum of 75.3% decolorization when used in conjunction with 2 mM hydrogen peroxide and 10 mg/L methyl orange at pH 3.0 for 40 min at 40 °C (Chiong et al. 2016). Three lipases—Lipase I, Lipase II, and Lipase III—were utilized by Meng et al. (2017) to hydrolyze vegetable oil, animal fat, and floating grease. Based on the hydrolysis settings of 40–50 °C, 1000–1500µL lipase inoculum, and 24 h, the results demonstrated that Lipase-I and Lipase-II efficiently liberated long-chain fatty acids from these pollutants. Lipase-III was shown to have a comparatively low rate of hydrolysis, indicating the significance of enzyme sources and types in wastewater treatment. Enzymes are vital to life activity, and every living cell can manufacture enzymes for biocatalysts. Some typical enzymes used and the contaminants they target are displayed in Table 1.1. Studies using animals as enzyme sources for wastewater treatment are few; however, they may serve as a source of lipase (Ning et al. 2016). Several plants are significant suppliers of enzymes for wastewater treatment, particularly

Table 1.1 Application of enzyme in the degradation of pollutants for application in wastewater treatment

Enzyme	Pollutant targeted	Wastewater type	References
Peroxidase	Phenols	Textile, leather	Garg et al. (2020)
Peroxidase	Dye	Textile	Darwish et al. (2019)
Peroxidase	Methyl orange, phenols	Textile, leather	Bilal et al. (2018)
Peroxidase	Methyl orange dye	Synthetic	Chiong et al. (2016)
Lignin peroxidase	Phenanthrene, benzo(a)pyrene, Dibutyl phthalate, tetracycline	Synthetic	Guo et al. (2019)
Lipase	Oil	Oily	Patel et al. (2020)
Lipase	Grease	Pig manure	Meng et al. (2017)
Lipase	Tributyrin	Synthetic	Jurado et al. (2008)
Laccase	Phenol, SMX and IPN, CBZ, chlorpyrifos, BPA	Synthetic	Fathali et al. (2019)
Laccase	Phenolic compounds	Palm oil mill	Theerachat et al. (2017)
Glucose oxidase	Indigo carmine in the presence of UV light	Textile	Atacan et al. (2019)

peroxidases (Ely et al. 2017; Garg et al. 2020). Enzymes from bacteria and fungi have been used extensively in wastewater treatment. For example, hydrolytic bacteria can create cellulase, amylase, lipase, and protease enzymes that can break down complex molecules found in wastewater, such as protein, lipids, carbohydrates, and so on (Liew et al. 2020). Triglycerides in palm oil mill effluent were successfully treated using exocellular lipase derived from Candida rugosa. Chlorpyrifos was broken down using laccase from Trametes versicolor (Das et al. 2017). A single strain is not the only one that can produce enzymes. Extracellular enzymes isolated from microbial communities in wastewater have been found to have superior catalytic capacity than intracellular enzymes, particularly when it comes to the biotransformation of antibiotics.

In 120 h, the undiluted palm oil mill effluent containing 98.5% triglycerides could be eliminated by the crude exocellular enzyme preparations of two co-cultured yeasts, Candida rugosa and Yarrowia lipolytica rM-4A (Theerachat et al. 2017). Though more research is still needed to identify the proper enzymes for a specific wastewater treatment, plants, bacteria, and fungi are significant suppliers of enzymes for pollutants in wastewater removal.

1.3   Performance of Enzyme-Driven Wastewater Pollutants Biodegradation

1.3.1   Oil and Grease

There is an enormous surplus of meat and edible oil due to the world's unchecked population expansion and resource consumption, not to mention the trash that results from it. According to one estimate, the value of the world's oil market was estimated to be US$83.4 billion in 2015 and would rise to US$130.3 billion by 2024. 2018 saw the production of about 203.8 million metric tonnes of edible vegetable oils and a consumption of 197.3 million metric tonnes (Luo et al. 2019). Additionally, during the past few decades, there has been a sharp rise in meat, slaughterhouses, and animal production (Cheng et al. 2020). The expanding meat and edible oil industries produce a significant volume of wastewater, including oil and grease (Lee et al. 2019). These wastewaters have a significant organic and inorganic load which, if released without adequate treatment, may impact aquatic habitats (Lee et al. 2019), even surface water (Ma et al. 2015) and will only contribute to global pollution.

The most often utilized enzymes in wastewater treatment that contain oil and grease are lipases, which have been used commercially. Lipase can break down oil and grease into simpler free fatty acids and glycerol (Cheng et al. 2020). Lipase has been observed to perform well in the hydrolysis of oil and grease. For instance, a commercial lipase was utilized to decompose the floating fatty wastes from the food processing sectors that deal with dairy and meat. Findings indicated that when the starting pH was changed to 7.0, this commercial lipase could encourage the release of long-chain fatty acids, particularly unsaturated acids (Pascale et al. 2019).

1.4   Pharmaceuticals

In the aquatic environment, pharmaceuticals are a prevalent organic micropollutant identified in recent decades. More than 200 distinct pharmacologically active chemicals have been found in water bodies as a result of widespread pharmaceutical use. Common medications found in wastewater include anti-infectives, cardiovascular drugs, central nervous system agents, metabolic agents, hormones, and more (Varga et al. 2019). Currently, traditional activated sludge systems do not remove enough medicines. Only four out of 35 medicines were found to have undergone above 90% removal when activated sewage sludge from municipal wastewater treatment plants was used (Joss et al. 2006).

In contrast, 17 compounds were found to have undergone less than 50% removal. According to a different study, wastewater treatment plants could only passively remove 20–90% of antibiotic residues. It is challenging for natural degradation to occur for antibiotics such as cephalosporins (Guo and Chen 2015), tetracyclines, and fluoroquinolones (Becker et al. 2016). Pharmaceuticals are common because leftover medications from treated wastewater are continuously released into the aquatic environment.

According to Caban et al. (2016), they have been found in drinking water, on the ground, and the surface. Even if medicines are present in wastewater released at low amounts, their bioaccumulation is at most negligible. In freshwater environments, concentrations of ciprofloxacin, norfloxacin, and oxytetracycline up to 6.5, 0.52, and 0.71 mg/L, respectively, were noted (Hughes et al. 2013). While some medications, such as DCF and CBZ, are ecotoxic (Lonappan et al. 2016), most medications' effects in aquatic environments are still primarily unclear. Because of their potential for bioaccumulation and ecotoxicology, inadequate or inefficiently treated medicines constitute a severe hazard to aquatic habitats. Oxidoreductases are efficient enzymes for eliminating pharmaceuticals, and some studies have demonstrated the viability of enzymatic procedures for pharmaceuticals involving wastewater treatment. When immobilized, chloroperoxidase (CPO) was employed by Song et al. (2019) to treat wastewater containing the antibiotics levofloxacin and rifampin; for example, over 88% of the antibiotics were destroyed in 30 min at an enzyme concentration of 20μg/mL. At 35 °C, pH 6, 60 U/L laccase concentration, and 18μM mediator concentration, a laccase and redox mediator system removed 95% of CBZ (Naghdi et al. 2018a). It is important to remember that the enzymatic process has the potential to produce hazardous byproducts from the conversion of medications like DCF (Li et al. 2017a, b) and CBZ (Naghdi et al. 2018a). Sadly, over 3000 distinct compounds have been utilized as components in medications, and because the biodegradation pathways of most pharmaceuticals are complicated and poorly understood, little is known about the environmental impacts of most pharmaceuticals (Stadlmair et al. 2018). The complex structures of pharmaceuticals hamper the use of enzyme-based medications that contain wastewater treatment. Thus, additional research is required to understand the mechanisms and pathways underlying the transformation of medicines based on enzymes.

1.5 Pesticides

Due to their high residue rate, limited utilization, and significant waste, pesticides seriously threaten human health and the world's aquatic ecosystems (Maier et al. 2015). Pesticides are highly toxic, and their residues will find their way into water bodies through percolation and runoff (Pal et al. 2014). Once there, they will degrade the water's quality, have an adverse effect on the biological, regulatory, and metabolic functions of aquatic life and upset the delicate balance of marine ecosystems (Moraes et al. 2007). Since they can function as cholinesterase inhibitors, organophosphorus and carbamate insecticides are widely used (Jiang et al. 2019), substantially harming the environment and animal health. There have been reports of other toxicities with potentially fatal consequences that are unrelated to cholinesterase. Certain highly hazardous pesticides, such as carbofuran, triazophos, and methamidophos, are prohibited in their usage, although demand for them remains high, particularly in developing nations. Necessary enzymes involved in the breakdown of pesticides have also been documented. For instance, in 12 h at pH 7 and 60 °C, the laccase immobilized on magnetic iron nanoparticles demonstrated a more than 99% clearance of chlorpyrifos (Das et al. 2017). The immobilized CPO could break down Isoproturon (IPN) at a concentration of 26.7 mol/L in under 10 min (Fan et al. 2018). Organophosphate degrading enzyme A (OpdA) could be covalently immobilized on nonwoven polyester textiles to break down small amounts of organophosphate insecticides efficiently (Gao et al. 2014). There have been reports of other enzymes that can decontaminate pesticides, such as organophosphorus acid anhydrolase (OPAA) and organophosphorus hydrolases (Ophs). In addition to destroying pesticides based on mono-enzymes, a successful outcome may also be obtained through the synergistic transformation of numerous enzymes. Acetylcholinesterase could convert acetylcholine chloride (Ach) into choline in a double-enzyme reaction, and choline oxidase (CHO) may further break down choline into betaine and hydrogen peroxide. A safe enzymatic method can be achieved by mixing numerous enzymes, as many enzymatic pesticide TPs are not environmentally benign.

1.6 Phenols and Phenolic Compounds

Phenols and phenolic compounds are essential industrial materials used in the textile, paper, plastic, pharmaceutical, and other sectors (Jun et al. 2019). These substances seriously threaten the environment even at low concentrations since they affect aquatic life and humans. Currently, research is being done on enzyme-based phenol and phenolic compound elimination. After 40 min at 35 °C and pH 4.5, an immobilized laccase could remove 0.4 mM of phenol (Fathali et al. 2019). Because their active sites—the hem cofactor and the redox-activated cysteine/selenocysteine residues—are easily accessible, peroxidases are efficient enzymes for converting phenolic substances (Garg et al. 2020). Studies at room temperature and pH 7.0 revealed that immobilized peroxidases could eliminate 99.9% of 1 mM phenol after 5 h of treatment (Wang et al. 2016). Similarly, Rezvani et al. (2015) used immobilized soybean seed coat powder as soybean peroxidase in a packed bed bioreactor to remove phenol from refinery wastewater. Under ideal circumstances, which included one mM starting phenol, 56 °C, 14 mM H_2O_2 concentration, and 5.5 mL/min flow rate, up to 97% phenol elimination was seen.

1.7 Plastics

Plastics are widely used in daily life, particularly in packaging, which not only makes products more convenient but also makes them non-biodegradable. The polymers that are often utilized are polyethylene (PE), polyamide (PA), polyethylene terephthalate (PET), polystyrene (PS), polyvinylchloride (PVC), polyurethanes (PUR), polypropylene (PP) (Danso et al. 2019). However, the fossil hydrocarbons from which polymers like ethylene and propylene are made are resistant to biodegradation. As a result, they will accumulate in the environment instead of breaking down (Geyer et al. 2017). To make matters worse, plastics are produced on a large scale with a low recycling rate. Just 9% of the more than 6.3 billion tonnes of plastic produced in 2015 were recycled (Geyer et al. 2017). Because of their toxicity and tendency to accumulate, residual plastics have a negative environmental impact (Zurier and Goddard 2021). In aquatic ecosystems, plastic is pervasive, particularly in the world's oceans (Geyer et al. 2017). It is also found in drinking water (Mintenig et al. 2019). Plastics resist biodegradation; however, certain enzymes, including tannase, MHETase, and PET hydrolase, can break them down (Danso et al. 2019). Unfortunately, the breakdown of plastics using enzymes is not very effective. PETase from Ideonalla sakaiensi (IsPETase), an effective and selective depolymerase, will require weeks to hydrolyze PET fully (Zurier and Goddard 2021).

Recent research has shown remarkable results in raising PETase's thermostability and enzyme activity (Son et al. 2019). While enzymes involved in the breakdown of PA and PET have been identified, little is known about the hydrolase enzymes of PVC, PP, PE, and PUR or potential pathways (Danso et al. 2019). However, since numerous bacteria have been discovered that are capable of degrading plastics, the degradation of these polymers by enzymes is feasible (Danso et al. 2019). While enzyme-based plastic degradation is a viable and promising approach, more investigation is required to optimize the diversity and efficiency of the enzymes involved to effectively remove polymers that become entangled in wastewater treatment processes.

1.8 Free Enzyme Systems

There are two types of enzymes in free enzyme systems: crude and refined. Two kinds of enzymes accomplished pollution removal (Zhang and Geissen 2010). However, since procedures like membrane separation, size exclusion chromatography, and column chromatography are needed for the purification of enzymes, purified enzymes are pricey (Zhu et al. 2013). Because crude enzymes contain mediators, purified enzymes were sometimes less effective than oil enzymes at eliminating wastewater contaminants (Tran et al. 2010). Naproxen (NPX) has been reported to be removed by crude laccase isolated from Trametes versicolor (Tran et al. 2010). Still, commercial laccase derived from Myceliophthora thermophila (Lloret et al. 2010) was only able to remove 60% of the drug. That does not imply, however, that unpurified enzymes are better than pure enzymes. More pollution will result from the mediators and unused nutrients in crude enzymes (Becker et al. 2016). Since enzymes are costly, these systems are impractical for treating wastewater due to the high demand for enzymes that cannot be recycled. Any leftover enzymes should be eliminated after the treatment process. If free enzymes are to be used for wastewater treatment, a great deal of work has to be done to address the issues of cost and contamination.

1.9 Immobilized Enzyme Systems

An often-used method to get around the drawbacks of using enzymes as biocatalysts for color treatment is immobilization. The attachment of soluble enzymes to a supporting matrix that causes the associated enzyme to lose some or all of its mobility is known as enzyme immobilization. The enzyme is immobilized and acquires improved characteristics, but strictly case-by-case. Stability under varied denaturing conditions, pH tolerance, functional stability, ease of enzyme separation and recovery, reusability, and improved catalytic efficiency are only some of the improved features.

Enzyme stabilization and prevention of dissociation-related inactivation are other benefits of mechanical stiffening of the enzymes through immobilization.

Similarly, the immobilized ginger peroxidase could achieve higher stability across a broader range of pH, higher reusability, and higher thermostability. On the contrary, the immobilization of enzymes might lead to decreased catalytic activity compared to free enzymes. Chagas and co-workers immobilized soybean peroxidase on chitosan beads using glutaraldehyde as a cross-linker for phenolic degradation. However, the enzyme activity became more thermally dependent after immobilization, which the work attributed to destroying the enzyme's active sites. This was due to the significant expansion coefficient of the chitosan support. In sum, enzyme immobilization can either increase or decrease an enzyme's parameter performance; changes in each parameter can only be determined by research. When appropriately used, enzyme immobilization can produce the required qualities for enzyme application.

Entrapment, adsorption, covalent bonding, and cross-linking are the most widely utilized physical or chemical methods for immobilizing enzymes. Physical immobilization techniques like entrapment and adsorption depend on weak interactions, including ion binding, van der Waals forces, and the physical restraint of the enzyme inside the support. Since the enzyme's natural structure is preserved, the physical immobilization approach has the advantage of maintaining the enzyme's catalytic activity. Adsorption has the most scalable mechanism, highest efficiency, and lowest cost of any known method for carrying out enzyme immobilization in the market today. The procedure entails submerging the support in an enzyme solution for a predetermined amount of time to immobilize the enzyme on the support. It is, however, readily disturbed due to the weak, non-specific forces holding the enzyme and support together. A method of immobilization called entrapment is employed to counteract the tendency of enzymes to aggregate, which reduces their catalytic activity. This technique of immobilizing enzymes involves gel/fiber entrapping, metal-organic frameworks (MOF), or microencapsulation. This approach of immobilizing the enzymes keeps the enzyme on the support while allowing the substrates to flow through the networks and catalyze the reaction. As a result, there is less enzyme leakage and an improvement in the enzyme's mechanical stability. Enzyme conformations remain unchanged because covalent bonds do not form between the enzyme and support. This ensures high catalytic activity.

However, the high diffusion barrier associated with entrapment hinders macro-molecular substrates from navigating the network of enzymes, which is a drawback. Creating covalent bonds between the enzymes and supporting them using bio-functional reagents like glutaraldehyde, covalent attachment, and cross-linking enzymes, on the other hand, are chemical immobilization techniques. The enzyme is slightly altered, which may affect its activity, but this is made up for by its stiffer structure and stronger chemical interactions. As a result, there is still a constant need to create a fresh immobilization technique to get beyond the constraints.

Ge and colleagues have recently identified an elegant immobilization approach by generating hybrid organic-inorganic nanoflowers using proteins as the organic component and copper (II) ions as the inorganic component (Ge et al. 2014). There were three main steps in the immobilization procedure. The paramount crystals were created in the first stage by nucleation. Expanding the biomolecules and the agglomerates of the paramount crystals constituted the second step. The last stage involved anisotropic growth, resulting in a structure resembling a branched flower. This completed the structure of the nanoflower, which contains the encapsulated enzyme. The coordination of Cu^{2+} and protein is the primary mechanism responsible for the production of the nanoflowers.

1.10   Enzyme Incorporated Nanotechnology

Numerous applications, including enzyme prodrug treatment, antibacterial agent, glucose biosensor and lactose hydrolysis have been investigated and employed with immobilized enzymes on nanoscale support. Even though other reviews have examined their performance in these various sectors, there has yet to be a thorough evaluation of the effectiveness of immobilized enzymes on nano support for dye decolorization. This chapter will analyze the effectiveness of enzyme-incorporated nanotechnology in the context of dye decolorization. Compared to their free counterparts, enzymes immobilized on nano support using traditional immobilization techniques showed superior reusability, increased thermal stability, and high activity over a more comprehensive pH range. For instance, ginger peroxidase was immobilized by Ali and colleagues on an amino-functionalized silica-coated titanium dioxide nano-composite. In both acidic and alkali areas, the immobilized ginger peroxidase demonstrated a much greater decolorization activity.

In contrast, the free peroxidase kept less than 10% of its activity after 100 min at 60 °C, whereas the immobilized ginger peroxidase was able to stay over 80% of its activity. The study ascribed the excellent activity preservation to the nanosupport's defense against reaction byproduct inhibition. In a different work, it was found chitosan nanoparticles on glass beads immobilize laccase and decolorize congo red. At 90 °C, the free laccase completely lost its decolorization activity due to enzyme denaturation, whereas the immobilized laccase retained almost all of its decolorization activity. Laccase was immobilized on Fe_3O_4/SiO_2 nanoparticles by Wang and colleagues to decolorize Procion Red MX-5B. Excellent reusability was demonstrated by the immobilized enzyme, which at the tenth cycle maintained 79% of its initial activity. The build-up of dye degradation products, restrictions on mass transfer, and loss of magnetic carriers during magnetic separation from the reaction media were all blamed for the decline in activity.

With encouraging findings, some investigations have also advanced using the recently developed hybrid nanoflower immobilization technique for dye decolorization. To remove crystal violet and neutral red, for example, Li and colleagues immobilized laccase in a self-assembly hybrid nano-composite made of carbon nanotubes, graphene oxide, and copper phosphate. Whereas the free laccase could only remove less than 20% of the dye, the hybrid nano-composite demonstrated outstanding dye removal at 100% elimination. The composite structure of carbon nanotubes and graphene oxide was shown to be responsible for the more significant activity of the immobilized laccase. This structure improved electron transfer efficiency and ensured a good affinity between the laccase and substrates. The immobilized laccase retained almost 80% of its initial activity after 6 days at room temperature, but the free enzyme only retained about 30%. In addition, the hybrid nanoflower outperformed the free enzyme regarding stability and reusability. These investigations have produced encouraging findings, indicating that most enzyme-incorporated nanotechnology can efficiently decolorize dye in a brief amount of time. Consequently, there is great promise in applying these biocatalysts to treat wastewater containing industrial dyes.

1.11 Enzyme Mediators-Related Systems

Kim and Nicell (2006) describe enzyme mediators as low-weight, stable substances that have the potential to function as an electron shutter between substrates and enzymes. By generating highly reactive radicals, mediators may increase the variety of enzyme substrates (Naghdi et al. 2018b). Additionally, these tiny molecules enhance the effectiveness of the enzymatic process and remove the targeted contaminants with greater efficiency (Munk et al. 2018). When the redox potential of the enzyme was greater than that of the substrate, mediators typically helped to increase the oxidation process (Madhavi and Lele 2009). There are two categories of enzyme mediators: natural mediators and synthetic mediators. While ABTS, HBT, and violuric acid (VLA) are synthetic mediators, syringaldehyde (SA), AS, VA, acetovanillone, methyl vanillate, and 4-hydroxybenzyl alcohol (HBA) are natural mediators. They have the power to enhance the enzymatic process and enzyme activity.

Orange G dye decolorization was achieved by Bibi et al. (2019) using laccase; the results indicated that the mediator might enhance both laccase activity and orange G dye removal. At 156.78 IU/mL, 61.45% of the orange G dye was eliminated; in the laccase-mediator system, however, 85% of the orange G dye was decolorized, and up to 97% of the dye was removed, all at a laccase activity of 279.27 IU/mL. Less than 30% of CBZ was broken down by free laccase after 24 h because it is not a reactive substrate for laccase; however, more than 82% of CBZ was removed when mediator ABTS was present (Naghdi et al. 2018a). When tetracycline was removed with horseradish peroxidase, the maximum amount of tetracycline degradation was approximately 53% in a pure enzyme system. In contrast, over 95% of the tetracycline was removed in a system with a mediator added, and the reaction time for these results decreased from 1 h to 0.5 h (Leng et al. 2020). Becker et al. (2016) used immobilized laccase to biodegrade 38 antibiotics. According to the findings, laccase in the absence of the mediator did not significantly reduce the number of antibiotics; however, when mediator SA was added, 32 out of 38 antibiotics exhibited more significant than 50% clearance.

The concentration of mediators significantly influences the elimination of contaminants. According to Naghdi et al. (2018a), increasing the mediator concentration from 6μM to 14μM enhanced the elimination of CBZ from 47% to 95%. Mediators may increase the effectiveness of pollutant removal; nevertheless, these methods are not always the best. For example, Ji et al. (2016) employed laccase to remove CBZ, and the results revealed that p-coumaric acid (PCA) as a mediator could remove 60% in 96 h, while a membrane hybrid reactor with immobilized laccase was able to remove 71% of the CBZ. Additionally, different mediators may affect a given response differently, resulting in varying TP toxicity (Ashe et al. 2016). The mixing of mediators could also bring on inadequate performance. According to reports, adding mediators SA and VLA separately may increase the pace at which TrOCs are broken down by laccase. Still, when combined with other medications, such as CBZ, IBF, and primidone, it may slow down the breakdown of those medications. It is important to note that mediators can alter the enzymatic process mechanism, producing distinct TPs and making it more challenging to regulate the enzymatic process (Varga et al. 2019). According to one study, adding mediator SA would make the effluent more hazardous when laccase was used to cure TrOCs. It is unclear if the harmful enzymatic TPs or the interaction between laccase and mediator caused this event, even though it is connected to the dose of SA (Nguyen et al. 2016).

1.12 Challenges and Strategies

1.12.1 Implementation at the Industrial Scale

With millions of people still without access to this daily essential, the demand for clean water is at an all-time high as the world's population continues to rise. However, supplying clean water to regions where industrial dye spills occur daily is still challenging. Because established and conventional dye wastewater treatment systems are frequently expensive and inefficient, irresponsible enterprises engage in the illicit dumping of untreated dye wastewater, damaging groundwater resources and water catchment areas. As a result, using enzymes to treat wastewater-containing dyes is a good option for effectively and economically removing the majority of the dye. As was previously said, many studies have been done to confirm that enzymes immobilized on nano support can treat wastewater containing a single dye, and promising findings have been found. Since much of the research is being done at the lab size, it is yet to be determined whether such systems can be implemented. A few studies have used enzyme-incorporated nanotechnology to remediate binary dye systems and mimicked industrial dye effluent. For instance, due partly to their remarkable regeneration and reusability qualities, graphene oxide (GO) based nanomaterials are excellent enzyme-incorporated adsorbents that are quickly commercialized. Recent studies carried out using magnetic GO-immobilized laccase (Chen et al. 2017), GO-immobilized peroxidase and GO-laccase nanoflowers have proven to degrade dyes from water efficiently in lab scale. Chen and co-workers discovered that 59.8% enzymatic activity was retained even after ten dye decolorization cycles in an aqueous solution containing mainly dye and phosphate buffer. However, it is difficult to determine the optimum conditions in the environmental bioremediation process as their effectiveness in treating actual dye-loaded wastewater from industries has yet to be evaluated. The actual performance of nanotechnology in treating dye wastewater could differ from the results produced in the laboratory. This indicates that, given the potential complexity of the composition of industrial dye discharge, current studies to determine the efficacy of enzyme-incorporated nanotechnology in actual dye wastewater treatment still need to be completed. Despite the potential of enzymes, converting lab-scale research into practical industrial applications would remain the main obstacle. Therefore, real-world industrial wastewater settings should be used to test enzyme-incorporated nanotechnology's efficacy in treating dye effluents. The focus of future research should be on handling actual industrial dye.

1.13   Understanding Enzymes Incorporated Nanotechnology

Thus far, works on dye decolorization utilizing enzyme-incorporated nanotechnology have made remarkable strides toward developing methods to increase the enzyme's catalytic activity. Numerous studies and analyses of the unique characteristics of enzymes and nanomaterials have already been conducted. Nevertheless, once integrated, it becomes more work to fully comprehend and do quantitative analysis on the combined system at such a small scale (Johnson et al. 2014). It is still challenging to isolate every variable experimentally to pinpoint the precise reason behind a system's enzymatic activity shift. For example, enzymes after immobilization can lose diffusional mobility while at the same time undergoing conformational changes, making it difficult to understand the actual cause of the change. Currently, there is no obvious solution to designing an experiment to separate these two effects of diffusional mobility and conformation changes. Also, it would be flawed to generalize the root cause of an enhancement or limitation of one system to another. Even the slightest change in the enzyme and support material combination can cause significant differences. For instance, Zhang, Luo, and Chen adsorbed catalase onto carbon nanotubes to determine the effect of the carbon nanotube surface properties on enzymatic activity (Zhang et al. 2013). The results suggested that conformational changes were responsible for the catalase structure's lower activity following immobilization. They used circular dichroism spectroscopy to look into laccase's post-immobilization secondary structure. They discovered no discernible conformational change in the immobilized laccase, even though the nanomaterial utilized was similar to that of Zhang, Luo, and Chen's earlier study. As a result, they concluded that substrate accessibility, rather than conformational change, was the cause of the loss in enzymatic activity. To remove acid yellow 42, it is reported immobilized ginger peroxidase on an amino-functionalized silica-coated titanium dioxide nanocomposite. Similarly, they used circular dichroism spectroscopy to show how the structure of ginger peroxidase changed conformationally and suggest that this could be the reason for the improved performance.

In summary, our investigations have demonstrated that various enzymes immobilized on various nanomaterials would result in systems that may differ from one another. Scientists still know very little about how enzyme-incorporated nanotechnology functions. As a result, further research should be conducted to better understand enzymes and nanotechnology by examining the effectiveness of specific variables. However, This does not rule out the prospect that improved methods for the treatment of dye wastewater and other industrial uses can be created with a thorough understanding of the underlying mechanisms involved in enhancing enzyme activity.

1.14 Conclusion and Future Perspectives

The breakdown of pollutants is greatly aided by enzymes, and enzymatic bioprocesses have proven to be the preferred method of treating wastewater. When it comes to the biodegradation of refractory pollutants, including grease, oil, medications, personal care items, insecticides, and industrial chemicals, enzymatic processes show promise. However, as enzymes can only break down complex molecules into simpler ones, this approach can only be used as a pretreatment and needs to be used in conjunction with other techniques to achieve complete therapy. There are now several barriers to the broader deployment of this technology, including the need for different enzyme system constructions, enzymatic TP monitoring, enzyme abilities augmentation, and investigations into natural wastewater and larger-scale applications. We suggest conducting additional research to understand better these systems' behavior, their viability in actual industrial settings, how they recover from reaction media, and how environmentally sustainable they are. The use of these biocatalysts in the treatment of dye wastewater has a bright future as long as the obstacles to its use are overcome to utilize these biocatalysts fully.

References

Al-Ansari MM, Modaressi K, Taylor KE, Bewtra JK, Biswas N (2010) Soybean peroxidase-catalyzed oxidative polymerization of phenols in coal-tar wastewater: comparison of additives. Environ Eng Sci 27(11):967–975

Alshabib M, Onaizi SA (2019) A review on phenolic wastewater remediation using homogeneous and heterogeneous enzymatic processes: current status and potential challenges. Sep Purif Technol 219:186–207

Ashe B, Nguyen LN, Hai FI, Lee D-J, van de Merwe JP, Leusch FDL, Price WE, Nghiem LD (2016) Impacts of redox-mediator type on trace organic contaminants degradation by laccase: degradation efficiency, laccase stability and effluent toxicity. Int Biodeterior Biodegrad 113: 169–176

Atacan K, Güy N, Çakar S, Özacar M (2019) Efficiency of glucose oxidase immobilized on tannin modified NiFe$_2$O$_4$ nanoparticles on decolorization of dye in the Fenton and photo-biocatalytic processes. J Photochem Photobiol A Chem 382:111935

Becker D, Giustina SVD, Rodriguez-Mozaz S, Schoevaart R, Barceló D, de Cazes M, Belleville M-P, Sanchez-Marcano J, de Gunzburg J, Couillerot O, Völker J, Oehlmann J, Wagner M (2016) Removal of antibiotics in wastewater by enzymatic treatment with fungal laccase—degradation of compounds does not always eliminate toxicity. Bioresour Technol 219:500–509

Bibi I, Javed S, Ata S, Majid F, Kamal S, Sultan M, Jilani K, Umair M, Khan MI, Iqbal M (2019) Biodegradation of synthetic orange G dye by Plearotus sojar-caju with Punica granatum peal as natural mediator. Biocatal Agric Biotechnol 22:101420

Bilal M, Rasheed T, Iqbal HMN, Hu H, Wang W, Zhang X (2018) Horseradish peroxidase immobilization by copolymerization into cross-linked polyacrylamide gel and its dye degradation and detoxification potential. Int J Biol Macromol 113:983–990

Caban M, Bialk-Bielinska A, Stepnowski P, Kumirska J (2016) Current issues in pharmaceutical residues in drinking water. Curr Anal Chem 12(3):249–257

Chen J, Leng J, Yang X, Liao L, Liu L, Xiao A (2017) Enhanced performance of magnetic graphene oxide-immobilized laccase and its application for the decolorization of dyes. Molecules 22(2). https://doi.org/10.3390/molecules22020221

Cheng D, Liu Y, Ngo HH, Guo W, Chang SW, Nguyen DD, Zhang S, Luo G, Liu Y (2020) A review on application of enzymatic bioprocesses in animal wastewater and manure treatment. Bioresour Technol 313:123683

Chiong T, Sie Yon JL, Lek Z, Boon Y, Danquah M (2016) Enzymatic treatment of methyl orange dye in synthetic wastewater by plant-based peroxidase enzymes. J Environ Chem Eng 4(2): 2500–2509

Cummins I, Landrum M, Steel PG, Edwards R (2007) Structure activity studies with xenobiotic substrates using carboxylesterases isolated from Arabidopsis thaliana. Phytochemistry 68(6): 811–818

Daâssi D, Prieto A, Zouari-Mechichi H, Martínez MJ, Nasri M, Mechichi T (2016) Degradation of bisphenol A by different fungal laccases and identification of its degradation products. Int Biodeterior Biodegrad 110:181–188

Danso D, Chow J, Streit WR (2019) Plastics: environmental and biotechnological perspectives on microbial degradation. Appl Environ Microbiol 85:e01095–e01119

Darwish AAA, Rashad M, AL-Aoh HA (2019) Methyl orange adsorption comparison on nanoparticles: isotherm, kinetics, and thermodynamic studies. Dyes Pigm 160:563–571

Das A, Singh J, Yogalakshmi KN (2017) Laccase immobilized magnetic iron nanoparticles: fabrication and its performance evaluation in chlorpyrifos degradation. Int Biodeterior Biodegrad 117:183–189

Ely C, de Lourdes Borba Magalhães M, Soares CHL, Skoronski E (2017) Optimization of phenol removal from biorefinery effluent using horseradish peroxidase. J Environ Eng 143(12): 04017075

Fan X, Hu M, Li S, Zhai Q, Wang F, Jiang Y (2018) Charge controlled immobilization of chloroperoxidase on both inner/outer wall of NHT: improved stability and catalytic performance in the degradation of pesticide. Appl Clay Sci 163:92–99. https://doi.org/10.1016/j.clay.2018. 07.016

Fathali Z, Rezaei S, Faramarzi MA, Habibi-Rezaei M (2019) Catalytic phenol removal using entrapped cross-linked laccase aggregates. Int J Biol Macromol 122:359–366

Gao Y, Truong YB, Cacioli P, Butler P, Kyratzis IL (2014) Bioremediation of pesticide contaminated water using an organophosphate degrading enzyme immobilized on nonwoven polyester textiles. Enzym Microb Technol 54:38–44

Garg S, Kumar P, Singh S, Yadav A, Dum'ee LF, Sharma RS, Mishra V (2020) Prosopis juliflora peroxidases for phenol remediation from industrial wastewater — an innovative practice for environmental sustainability. Environ Technol Innov 19:100865

Ge Y, Li Z, Xiao D, Xiong P, Ye N (2014) Sulfonated multi-walled carbon nanotubes for the removal of copper (II) from aqueous solutions. J Ind Eng Chem 20(4):1765–1771

Geyer R, Jambeck J, Law K (2017) Production, use, and fate of all plastics ever made. Sci Adv 3(7): e1700782

Guo R, Chen J (2015) Application of alga-activated sludge combined system (AASCS) as a novel treatment to remove cephalosporins. Chem Eng J 260:550–556

Guo J, Liu X, Zhang X, Wu J, Chai C, Ma D, Chen Q, Xiang D, Ge W (2019) Immobilized lignin peroxidase on Fe_3O_4@SiO_2@polydopamine nanoparticles for degradation of organic pollutants. Int J Biol Macromol 138:433–440. https://doi.org/10.1016/j.ijbiomac.2019.07.105

Hughes SR, Kay P, Brown LE (2013) Global synthesis and critical evaluation of pharmaceutical data sets collected from river systems. Environ Sci Technol 47(2):661–677

Ji C, Hou J, Wang K, Zhang Y, Chen V (2016) Biocatalytic degradation of carbamazepine with immobilized laccase-mediator membrane hybrid reactor. J Membr Sci 502:11–20

Jiang B, Zhang N, Xing Y, Lian L, Chen Y (2019) Microbial degradation of organophosphorus pesticides: novel degraders, kinetics, functional genes, and genotoxicity assessment. Environ Sci Pollut Res 26(21):21668–21681

Johnson BJ, Algar WR, Malanoski AP, Ancona MG, Medintz IL (2014) Understanding enzymatic acceleration at nanoparticle interfaces: approaches and challenges. Nano Today 9(1):102–131

Joss A, Zabczynski S, Gobel A, Hoffmann B, Loffler D, McArdell CS, Ternes TA, Thomsen A, Siegrist H (2006) Biological degradation of pharmaceuticals in municipal wastewater treatment: proposing a classification scheme. Water Res 40(8):1686–1696

Jun LY, Yon LS, Mubarak NM, Bing CH, Pan S, Danquah MK, Abdullah EC, Khalid M (2019) An overview of immobilized enzyme technologies for dye and phenolic removal from wastewater. J Environ Chem Eng 7(2):102961

Jurado E, Camacho F, Luzón G, Fernández-Serrano M, García-Román M (2008) Kinetics of the enzymatic hydrolysis of triglycerides in o/w emulsions. Biochem Eng J 40(3):473–484

Kim YJ, Nicell JA (2006) Laccase-catalysed oxidation of aqueous triclosan. J Chem Technol Biotechnol 81(8):1344–1352

Lee ZS, Chin SY, Lim JW, Witoon T, Cheng CK (2019) Treatment technologies of palm oil mill effluent (POME) and olive mill wastewater (OMW): a brief review. Environ Technol Innov 15: 100377

Leng Y, Bao J, Xiao H, Song D, Du J, Mohapatra S, Werner D, Wang J (2020) Transformation mechanisms of tetracycline by horseradish peroxidase with/without redox mediator ABTS for variable water chemistry. Chemosphere 258:127306

Li X, He Q, Li H, Gao X, Hu M, Li S, Zhai Q, Jiang Y, Wang X (2017a) Bioconversion of non-steroidal anti-inflammatory drugs diclofenac and naproxen by chloroperoxidase. Biochem Eng J 120:7–16

Li H, Hou J, Duan L, Ji C, Zhang Y, Chen V (2017b) Graphene oxide-enzyme hybrid nanoflowers for efficient water soluble dye removal. J Hazard Mater 338:93–101. https://doi.org/10.1016/j. jhazmat.2017.05.014

Liew YX, Chan YJ, Manickam S, Chong MF, Chong S, Tiong TJ, Lim JW, Pan GT (2020) Enzymatic pretreatment to enhance anaerobic bioconversion of high strength wastewater to biogas: a review. Sci Total Environ 713:136373

Lloret L, Eibes G, Lú-Chau TA, Moreira MT, Feijoo G, Lema JM (2010) Laccase-catalyzed degradation of anti-inflammatories and estrogens. Biochem Eng J 51:124–131

Lonappan L, Brar SK, Das RK, Verma M, Surampalli RY (2016) Diclofenac and its transformation products: environmental occurrence and toxicity – a review. Environ Int 96:127–138

Luo Q, Liu ZH, Yin H, Dang Z, Liu Y (2019) Global review of phthalates in edible oil: an emerging and nonnegligible exposure source to human. Sci Total Environ 704:135369

Ma Z, Lei T, Ji X, Gao X, Gao C (2015) It submerged membrane bioreactor forVegetable oil wastewater treatment. Chem Eng Technol 38(1):101–109

Madhavi V, Lele SS (2009) Laccase: properties and applications. Bioresources 4(4):1694–1717

Maier D, Blaha L, Giesy JP, Henneberg A, Köhler H-R, Kuch B, Osterauer R, Peschke K, Richter D, Scheurer M (2015) Biological plausibility as a tool to associate analytical data for micropollutants and effect potentials in wastewater, surface water, and sediments with effects in fishes. Water Res 72:127–144. https://doi.org/10.1016/j.watres.2014.08.050

Meng Y, Luan F, Yuan H, Chen X, Li X (2017) Enhancing anaerobic digestion performance of crude lipid in food waste by enzymatic pretreatment. Bioresour Technol 224:48–55

Mintenig SM, Loder MGJ, Primpke S, Gerdts G (2019) Low numbers of microplastics detected in drinking water from ground water sources. Sci Total Environ 648:631–635

Moraes BS, Loro VL, Glusczak L, Pretto A, Menezes C, Marchezan E, de Oliveira Machado S (2007) Effects of four rice herbicides on some metabolic and toxicology parameters of teleost fish (Leporinus obtusidens). Chemosphere 68(8):1597–1601

Munk L, Andersen ML, Meyer AS (2018) Influence of mediators on laccase catalyzed radical formation in lignin. Enzym Microb Technol 116:48–56

Naghdi M, Taheran M, Brar SK, Kermanshahi-pour A, Verma M, Surampalli RY (2018a) Bio-transformation of carbamazepine by laccase-mediator system: kinetics, by-products and toxicity assessment. Process Biochem 67:147–154

Naghdi M, Taheran M, Brar SK, Kermanshahi-Pour A, Verma M, Surampalli RY (2018b) Removal of pharmaceutical compounds in water and wastewater using fungal oxidoreductase enzymes. Environ Pollut 234:190–213

Nguyen LN, van de Merwe JP, Hai FI, Leusch FDL, Kang J, Price WE, Roddick F, Magram SF, Nghiem LD (2016) Laccase–syringaldehyde-mediated degradation of trace organic contaminants in an enzymatic membrane reactor: removal efficiency and effluent toxicity. Bioresour Technol 200:477–484. https://doi.org/10.1016/j.biortech.2015.10.054

Ning Z, Ji J, He Y, Huang Y, Liu G, Chen C (2016) Effect of lipase hydrolysis on biomethane production from swine slaughterhouse waste in China. Energy Fuel 30(9):7326–7330

Pal A, He Y, Jekel M, Reinhard M, Gin KYH (2014) Emerging contaminants of public health significance as water quality indicator compounds in the urban water cycle. Environ Int 71:46–62

Pascale NC, Chastinet JJ, Bila DM, Sant Anna GL, Quiterio SL, Vendramel SMR (2019) Enzymatic hydrolysis of floatable fatty wastes from dairy and meat foodprocessing industries and further anaerobic digestion. Water Sci Technol 79(5):985–992

Patel H, Ray S, Patel A, Patel K, Trivedi U (2020) Enhanced lipase production from organic solvent tolerant Pseudomonas aeruginosa UKHL1 and its application in oily waste-water treatment. Biocatal Agric Biotechnol:101731. https://doi.org/10.1016/j.bcab.2020.101731

Rajesh A, Jai GS, Bhanumati S, Krishan K, Pradip N, Saroj K (2017) A simple and efficient method for removal of phenolic contaminants in wastewater using covalent immobilized horseradish peroxidase. J Mater Sci Eng B 7(1):27–38

Rezvani Romeh AA, Hendawi MY (2014) Bioremediation of certain organophosphorus pesticides by two biofertilizers, Paenibacillus(Bacillus) polymyxa (Prazmowski) and Azospirillum lipoferum (Beijerinck). J Agric Sci Tech-IRAN 16(2):265–276

Rezvani F, Azargoshasb H, Jamialahmadi O, Hashemi-Najafabadi S, Mousavi SM, Shojaosadati SA (2015) Experimental study and CFD simulation of phenol removal by immobilization of soybean seed coat in a packed-bed bioreactor. Biochem Eng J 101:32–43

Sauvé S, Desrosiers M (2014) A review of what is an emerging contaminant. Chem Cent J 8(1):1–7

Son HJ, Kim YS, Kim SG, Hwang YD, Hwang IS (2019) Insight into biodegradation of dissolved organic matter fractions using LC-OCD-OND in drinking water treatment processes. J Korean Soc Environ Eng 41(1):55–60

Song Y, Ding Y, Wang F, Chen Y, Jiang Y (2019) Construction of nano-composites by enzyme entrapped in mesoporous dendritic silica particles for efficient biocatalytic degradation of antibiotics in wastewater. Chem Eng J 375:121968. https://doi.org/10.1016/j.cej.2019.121968

Stadlmair LF, Letzel T, Drewes JE, Grassmann J (2018) Enzymes in removal of pharmaceuticals from wastewater: a critical review of challenges, applications and screening methods for their selection. Chemosphere 205:649–661

Teodosiu C, Gilca A-F, Barjoveanu G, Fiore S (2018) Emerging pollutants removal through advanced drinking water treatment: a review on processes and environmental performances assessment. J Clean Prod 197:1210–1221. https://doi.org/10.1016/j.jclepro.2018.06.247

Theerachat M, Tanapong P, Chulalaksananukul W (2017) The culture or co-culture of Candida rugosa and Yarrowia lipolytica strain rM-4A, or incubation with their crude extracellular lipase and laccase preparations, for the biodegradation of palm oil mill wastewater. Int Biodeterior Biodegradation 121:11–18. https://doi.org/10.1016/j.ibiod.2017.03.002

Tran NH, Urase T, Kusakabe O (2010) Biodegradation characteristics of pharmaceutical sustances by whole fungal culture Trametes versicolor and its laccase. J Water Environ Technol 8:125–140

Varga B, Somogyi V, Meiczinger M, Kováts N, Domokos E (2019) Enzymatic treatment and subsequent toxicity of organic micropollutants using oxidoreductases - a review. J Clean Prod 221:306–322

Wang S, Fang H, Yi X, Xu Z, Xie X, Tang Q, Ou M, Xu X (2016) Oxidative removal of phenol by HRP–immobilized beads and its environmental toxicology assessment. Ecotoxicol Environ Saf 130:234–239

Zhang Y, Geissen SU (2010) In vitro degradation of carbamazepine and diclofenac by crude lignin peroxidase. J Hazard Mater 176(1–3):1089–1092

Zhang R, Zhang Y, Zhang Q et al (2013) Optical visualization of individual ultralong carbon nanotubes by chemical vapour deposition of titanium dioxide nanoparticles. Nat Commun 4:1727. https://doi.org/10.1038/ncomms2736

Zhu L, Gong L, Zhang Y, Wang R, Ge J, Liu Z, Zare RN (2013) Rapid detection of phenol using a membrane containing laccase nanoflowers. Chem Asian J 8:2358–2360. https://doi.org/10.1002/asia.201300020

Zurier HS, Goddard JM (2021) Biodegradation of microplastics in food and agriculture. Curr Opin Food Sci 37:37–44

Chapter 2
Enzymatic Degradation of Synthetic Plastics: New Insights

Abstract Since plastic has been used in most industries and sectors, including building and construction, packaging, electronics, healthcare, and transportation, pollution from plastic has become a global issue. Due to the ever-increasing strain this has placed on the ecosystem, many scientists and environmentalists are focusing on developing strategies to counteract this threat. Numerous investigations have been conducted to discover how to harness the undiscovered capacities of bacteria that can use plastics as their only source of carbon to naturally break down plastics since the discovery of microbial enzymes that can act on plastic in its natural environment and since it is a much faster and more effective method than others, enzymatic breakdown of plastics has been suggested to serve this goal. This study provides an overview of the use of enzymes from several sources to break down plastics and future research opportunities in this field.

Keywords Biodegradation · Enzymatic degradation · Synthetic plastics · Hydrolytic cleavage

2.1 Introduction

Most synthetic polymers used to make plastics are chains of carbon atoms with surrounding chains of hydrogen, oxygen, nitrogen, and sulfur (Shah et al. 2008). Since the 1950s, when plastics began, these polymers have been used in every aspect of our lives. Plastic production has increased significantly to fulfill the global demand for products and services. In the past 70 years, 8.3 billion tonnes of plastic have been created (Kamboj 2016). Most of this growth in plastic production is caused by the creation of single-use and throwaway plastics, which make up around half of all plastic items. Due to the significant rise in the disposal of plastics, which are mainly non-biodegradable and hence build up in the environment, our land and oceans have been polluted (Shimao 2001). Merely 21% of the total plastic generated worldwide is effectively managed.

In contrast, the remaining plastic finds its way into the environment as waste on land, where it accumulates and contaminates the surrounding ecosystem, or as ocean waste that contaminates water supplies. Every year, plastic waste dumped into the oceans kills millions of marine animals, including endangered species. According to Gomez et al. (2018), entanglements and ingestion are among the common causes of animal deaths in both water and on land.

When techniques such as incineration are employed to dispose of waste, the accumulation of plastic waste in the environment leads to chemical pollution. Chemicals used in the plastic production process, such as ethylene, vinyl chloride, terephthalic acid, and bisphenol, which are most frequently used in consumer-based plastic products, are detrimental to human health and can result in congenital disabilities, endocrine disorders, cancer, and a host of other conditions (Waring et al. 2018). People are currently eager to learn faster and more effective ways to deal with the massive buildup of this hazardous material because it is tough to stop producing these beneficial and widely used plastic products entirely. The solution is undoubtedly within the purview of our experts, who continuously strive to develop a biologically safe method of removing the plastic trash dumps that are rapidly piling up (Trevino et al. 2012). It has been found that some microbes develop enzymes that can break down these polymers (Tsushima and Matsushita 2010). According to Pometto et al. (1992), the enzymes are known to be members of the hydrolase family, including lipases, depolymerases, esterases, and PETases. These enzymes are known to break down the carbon backbone of many regularly used plastics. These enzymes, called hydrolytic enzymes because they work best in water, break down the plastic polymer into simpler monomeric units that are easier to modify in the environment and more readily assimilated by microorganisms as a carbon source. These products are subsequently broken down further into CO_2, H_2O, CH_4, and N_2 (Seo et al. 2009). By utilizing the degradation byproducts in other areas, the potential use of these metabolic byproducts of plastic degradation by microorganisms can be highly beneficial in reducing the current problem of plastic pollution and starting a recycling chain of the hazardous materials that pollute the environment and are hard to get rid of (Villain et al. 1995).

2.2 Types of Plastics Frequently Used

Because of their versatility and ability to be molded using a variety of chemicals, plastics have revolutionized every industry (Vona et al. 1965). These are made with different qualities, each tailored to a specific use (Pavia et al. 1988). The following describes some of the most widely used plastics for various applications and their primary uses in multiple industries.

2.2.1 Polyethylene (PE)

Polyethylene is a class of plastic that includes one of the most widely used varieties. It is a member of the thermoplastics class and is a polymer of ethylene monomers.

Since it started being produced commercially in the 1950s, millions of tonnes of it are produced annually and used in various applications and industries. Low-density and high-density polyethylene applications are entirely distinct; the former is utilized in carry bags, while the latter is employed in the building industry.

2.2.2   Polyethylene Terephthalate (PETE/PET)

Polythene, the most popular synthetic polymer for producing single-use, throwaway beverage bottles, is highly robust and complex terephthalate. It is a thermoplastic made by polymerizing terephthalic and ethylene acid by creating ester bonds. One of the primary causes of plastic pollution is single-use plastic products that are discarded on land or in the water. The toughness and strength of PET repeating units are attributed to their massive aromatic ring (Gross and Kalra 2002).

2.2.3   Polypropylene (PP)

One of the most used plastics, polypropylene, was first polymerized in 1951 and is made of propylene monomers (Carniel et al. 2016). This thermoplastic has become widely known in various industries, including the textile, automotive, and packaging sectors (Williams 1981). The ability of this plastic to copolymerize with other polymers to increase the diversity of products accessible to meet diverse needs is one of its main advantages (Rowe and Howard 2002). Being a thermoplastic, polypropylene can take on various shapes when heated up without losing its structural integrity. As a result, it has numerous applicable industrial uses (Nakkabi et al. 2015).

2.2.4   Polyvinyl Chloride (PVC)

Polyvinyl chloride ranks third globally in synthetic plastic material usage after polyethylene and polypropylene. Before the plasticizers are added, the substance has a whitish, brittle appearance. Two primary forms can be made: stiff and flexible. Rigid PVC is given mechanical qualities that are more flexible and smooth by adding plasticizers, such as phthalates. PVC is generally much more rigid, affordable, and has a far higher tensile strength than other plastics on the market. Nevertheless, burning it releases hydrogen chloride (HCl) vapors, which harm the health of those around them. For this reason, it is not advised in devices such as electrical wire insulators (Vona et al. 1965).

2.2.5   Polystyrene (PS)

A homopolymer made up of styrene repeating units is called polystyrene. Depending on its type, it can be either a thermoplastic or a thermosetting plastic. It is a translucent thermoplastic substance by nature. This plastic is a pioneer in the packaging sector since its firm shape, a foam material, is utilized in most applications. Because polystyrene has no smell and is non-toxic, it is an excellent material for food packing. However, because it is tossed and disposed of primarily, it is relatively inert and offers a significant environmental contamination risk (Tokiwa and Calabia 2004).

2.2.6   Polylactic Acid (PLA)

Since PLA plastic is made from renewable resources like sugar cane and maize starch, it differs from many other plastics. For this reason, it's called a bioplastic. It is biodegradable and shares characteristics with polyethylene, polypropylene, and polystyrene (Gross and Kalra 2002). Compared to other plastics that take over 500 years to disintegrate, natural breakdown usually occurs in a little 6 weeks to 24 months. It is among the safest plastics to use globally due to its cost-effective manufacture (Williams 1981).

2.2.7   Polybutylene Succinate (PBS)

The thermoplastic polybutylene succinate has characteristics similar to those of polypropylene. Because it naturally breaks down into carbon dioxide and water, it is a biodegradable plastic that can be utilized successfully. Succinic acid and 1–4-butanediol are used in its synthesis. It is more pliable than polylactic acid polymers and doesn't require plasticizers to form (Gross and Kalra 2002).

2.2.8   Polyurethane (PU)

Organic units connected by carbamate linkages are used to make polyurethane. They could be an ether or an ester. They don't dissolve in water. One of the most versatile types of plastic, it can be used to make sealants, elastomers, medical devices, flexible and rigid foam, and various other products (Nakkabi et al. 2015).

2.3 Enzymatic Degradation of Synthetic Plastics

2.3.1 Insights into the Enzymatic Cleavage of Polymers

Low degradation yields are typically the result of using live organisms as catalysts for solid substrates, which is a significant issue (Andler et al. 2020). Nowadays, the host systems for the production of enzymes, including cutinases, laccases, peroxidases, and esterases, are both bacteria and fungi. Al-Tammar et al. (2016) state that one of the main obstacles to the industrial use of enzymes for plastic breakdown is their instability in various environments, including high temperatures and harsh pH levels. Because of this, numerous researchers have looked for effective enzymes in various microbes to show how they might be used in industrial settings (Saravanan et al. 2021). Here, we showcase the most promising enzymes for breaking down plastic. The studies are divided into two groups based on the type of bonds that hold the monomers together for easier comprehension. PET and PUR belong to the second group, whereas plastics having a heteroatomic backbone, like PE, PP, PS, and PVC, belong to the first group.

2.3.2 Enzymatic Degradation of Plastics with a C–C Backbone

Plastics containing C–C bonds in their structure biodegrade poorly, mainly because they lack hydrolyzable functional groups. One of the most significant problems is the lack of physicochemical pretreatment procedures (Restrepo-Flórez et al. 2014). It has been suggested that natural polymers having C–C bond-degrading enzymes are excellent candidates for the biodegradation of polyolefins (Chen et al. 2020). This is the case for enzymes that break down lignin, the second most prevalent natural polymer in nature (after cellulose). Laccases and peroxidases are the primary enzymes that can degrade this class of polymers. Yoshida et al. (2016) initially described laccases (benzenediol: oxygen oxidoreductase; p-diphenol oxidase EC 1.10.3.2) as multicopper oxidase enzymes found in Chinese or Japanese lacquers (Rhus sp.). By transferring electrons from organic substrates to molecular oxygen, these enzymes, which are extensively distributed in nature, can catalyze the oxidation of various substrates, including phenolic substances, non-phenolic substrates, and environmental contaminants. Many microorganisms, including filamentous fungi, some plants, bacteria, and insects, create laccases in the natural world (Janusz et al. 2020). The organism determines the roles that laccases play. Plants include enzymes called laccases that break down lignin (Alessandra et al. 2010). According to Janusz et al. (2020), morphogenesis, fungal plant pathogen/host interactions, and stress defense are all impacted by fungal laccases. In addition to aiding in sporulation and pigmentation processes, bacterial laccases can provide resistance against oxidation and UV light (Alessandra et al. 2010; Janusz et al. 2020).

Because laccases are widely used in various biotechnological processes, including the bioremediation of industrial waste, pharmaceutical production, and textile bleaching, there is considerable interest in investigating laccases (Lecourt et al. 2021). Chen et al. (2020) have reported that the T1 site, which is one of the four Cu centers found in the X-ray structure of a laccase from Trametes trogii, is the location with the highest redox potential activity in laccases and is thought to be in charge of substrate oxidation. This mechanism, which would mainly function for phenolic lignin substrates, is theoretical. Though significant efforts have been undertaken, particularly toward their capacity to lower polymer molecular weights, few studies have been able to link microbial or fungal laccase with the ability to degrade polyethylene (Cowan et al. 2022). H_2O_2 is used as an electron-accepting co-substrate in oxidation-reduction reactions to promote the oxidation of organic and inorganic compounds by a group of oxidoreductases known as peroxidases (EC number 1.11.1.x) (Twala et al. 2020). The functions of this protein family are critical to industrial and biotechnological operations. Its application is typically associated with dye discoloration, bioremediation, bioenergy, and textile sector operations. Numerous animals, plants, and microbes, such as Bacillus subtilis, Trametes villosa, Phanerochaete chrysosporium, and P. tremellosa, are capable of producing these sets of enzymes (Amobonye et al. 2021). The two most often utilized peroxidases in C-C backbone plastic degradation processes are lignin peroxidases (LiPs EC 1.11.1.14) and manganese peroxidases (MnPs EC 1.11.1.13), among this broad category.

The active site of these enzymes contains FeIII, which reacts with H_2O_2 to generate a reactive compound. The complex is reduced by an electron donor—possibly a C–C polymer—to create a new molecule, which is then reduced once more to restore the original enzyme structure (Chen et al. 2020). The C–C bonds in lignin can be broken by liginolytic enzymes, although comparing these linkages to those found in synthetic polymers is challenging. Lignin's complicated structure offers an extensive range of C–C bonds, with energies ranging from 205 to 308 kJ mol^{-1}. According to Popov and Knyazev (2014), the PE C–C bond breakdown requires about 350 kJ mol^{-1}. Therefore, without physicochemical pre-treatments, it is doubtful that this enzyme will significantly break down synthetic polymers with C–C linkages. A recent study examined the breakage of C–C bonds in polymers like PS and PE using quantum mechanism computations. The findings imply that laccase and peroxidase enzymes may be able to extract a hydrogen anion in specific operating environments, which would prevent a hydride anion from being present at the Cβ location. According to Xu et al. (2019), this alteration would cause a considerable reduction of the C2-C3 bond, leading to bond cleavage.

According to the examined research, most enzymatic degradation procedures involving polymers with a C–C backbone fail to produce muscular degradation, mainly when physicochemical pretreatments are not applied at the initial degradation stage. The enzymes laccase and peroxidase attack the oligomers, or breakdown products, left over after polymer cleavage. For these polymers, the efficiency of enzymatic mechanisms mostly depends on whether a step of integrating oxygen or double bonds has been completed previously, which frequently requires the assistance of abiotic elements (Wei et al. 2020). The abiotic component thought to have the most significant impact on environmental plastic waste is photo-initiated oxidative breakdown. It forms free radicals, which are more challenging to produce in polymers with no double bonds, such as PE and PP. If radicals are created, they combine with ambient oxygen to produce radical peroxides (Singh and Sharma 2008). The result of this propagation stage is cross-linking or chain scission. This leads to the formation of carbonyl functional groups, such as aldehydes and ketones, a decrease in molecular weight, an increase in hydrophilicity, and a polymer that is more prone to fragmentation, all of which facilitate an enzymatic attack (Gewert et al. 2015). More proof is required to demonstrate that microbes and their enzymes, rather than outside sources, are the true origin of the polyolefin biodegradation processes.

2.3.3 Enzymatic Degradation of Plastics with Heteroatoms in the Main Chain

Ester or amide linkages in the class of polymers with heteroatomic backbones, which includes PET and PUR, are more hydrolyzable than C–C bonds (Wei and Zimmermann 2017). These polymers are vulnerable to hydrolytic attack, as Tang et al. (2021) explained, and the mechanisms involved include photooxidation, hydrolysis, and biodegradation. According to Chen et al. (2020), the primary enzymes known to cleave this class of polymers are hydrolases, lipases, esterases, proteases, and ureases. Cutin hydrolases, also known as cutinases (EC 3.1.1.74), are extracellular enzymes that can be induced and are present in bacteria and fungi (Al-Tammar et al. 2016). They can cleave the ester bonds in cutin. According to Al-Tammar et al. (2016), their hydrolytic activity can affect a variety of water-soluble esters, polymers, triglycerides, and synthetic fibers. The most prevalent PET hydrolytic enzymes are cutinases, which consistently fall within the serine hydrolase family alongside lipases and carboxylesterases. Cutinases are secreted by microorganisms that can break down plant material, hydrolyze cutin, and release cutin monomers. Cutinases are found in both bacteria and fungi, and they work in similar ways. Due to variations in sequence and structure, which most likely contribute to their higher thermostability, bacterial enzymes belong to a distinct cutinase subfamily (Chen et al. 2018). Actinobacteria have been shown to include PET-degrading enzymes from Thermobifida species (T. cellulolytic, T. alba, and T. halotolerant), Thermomonospora phylum (T. curvata), and Saccharomonospora (S. viridis). Due to the variations in their hydrophobic and electrostatic surface characteristics, the closely related cutinases Thc_CutI and Thc_Cut2 from T. cellulosilytica have been demonstrated to have different hydrolytic efficiencies on PET (Herrero Acero et al. 2013). Extracellular hydrolases Tcur1278 and Tcur0390 from T. curvata can break down PET and poly(ecaprolactone) and have catalytic and structural characteristics akin to cutinases from T. fusca and T. cellulolytic (Wei et al. 2014). Cut190, an enzyme like cutinase isolated from compost in Okayama, Japan, has been demonstrated to effectively catalyze surface hydrolysis and break down the inner block of PET (Kawabata et al. 2017). The breakdown of PET into bis(2-hydroxyethyl) terephthalate (BHET), mono(2-hydroxyethyl) terephthalate (MHET), and terephthalic acid (TPA) is catalyzed by the cutinase-like PET hydrolase, known as PETase (EC 3.1.1.101.). The hydrolase known as MHETase (EC 3.1.1.102) subsequently breaks down MHET to produce TPA, which the central metabolism can utilize through the TPA degradation pathway. Compared to previously identified cutinases, PETase has better hydrolytic activity and substrate selectivity towards PET despite sharing the greatest similarities with the PET hydrolase from T. fusca. Significant variations can be seen in the crystal structure of PETase, including an extra disulfide link (Chen et al. 2018). MHETase is not similar to any of the known MHET-degrading enzymes, which often also digest PET, and it does not demonstrate activity against PET. The TPA degradation gene clusters tphI and tphII of Comamonas sp. strain E6 are nearly identical to a complete gene cluster adjacent to the MHETase-encoding gene in the genome of I. sakaiensis. It is believed that the related enzymes protocatechuate (PCA) dehydrogenase and TPA 1,2-dioxygenase catalyze the TPA monomer's downstream metabolism (Yoshida et al. 2016). The PETase protein was extracted from I. sakaiensis and produced in an E. coli system by Liu et al. (2018). They discussed the substrate interaction of α/β hydrolase and the conservation of its catalytic machinery. In this work, several mutations were produced at particular locations within the PETase structure to enhance PET degradation via PETase activity: substrate binding pockets (W130, M132, W156, I180, Q90, S185, and S209), triad center residues (S131, D177, H208), and residues involved in proper active site folding (W68, Q153, R94, the N212). Consequently, the authors discovered that the mutations W68, Q153, R94, and N212 might be involved in a reduction in PETase's hydrolytic activity. Furthermore, the PETase protein mutations Q90, M132, and W156 may prevent PET from being recognized, which would lower the hydrolytic activity of the enzyme. Last but not least, according to Liu et al. (2018), mutations Y58A, W130A, W130H, and A180I raised PETase activity and produced various surface alterations in PET material. It has been reported that lipases (EC 3.1.13) act as catalysts in the hydrolysis of PUR and PET (Wilkes and Aristilde 2017). It was investigated how using various Pseudomonas strains to degrade a PUR of economic interest would affect the process. Two extracellular lipases, PueA and PueB, were found to be involved in the degradation process following biochemical and mutational investigations (Hung et al. 2016). Thermomyces lanuginosus lipase was thoroughly examined for PET surface modification. The material's hydrophilicity was increased due to this enzyme's effective hydrolysis of PET, which produced surface polar groups (Gricajeva et al. 2022). Furthermore, one of the most researched and adaptable lipases, lipase B from Candida antarctica (CALB), has shown remarkable catalytic efficiency in depolymerizing PET (Carniel et al. 2017). Esterases, in addition to lipases and cutinases, have been demonstrated to hydrolyze PET and PUR. A broad class of enzymes known as true esterases (EC 3.1.1.3 carboxyl ester hydrolases) usually hydrolyze the ester bonds of shorter-chain fatty acids compared to lipases. They have a broad substrate range and are known to help with various catabolic processes or the use of carbon sources (Gricajeva et al. 2022). The p-nitrobenzyl esterase BsEstB from B. subtilis (Ribitsch et al. 2017) and the serine esterase PmC from P. mendocina (Ronkvist et al. 2009), as well as esterases from T. alba and T. halotolerant, have all been used to degrade PET. Thh_Est from T. halotolerans and the thermoactive Est119 from T. alba were categorized as esterases (Hu et al. 2010; Ribitsch et al. 2012). Prior research suggested that lipases or cutinases were more effective at hydrolyzing the polyester surface than esterases were at breaking down polyesters (Ribitsch et al. 2012). It is hypothesized that esterase activity cleaves PUR. The urethanase bacterial enzyme hydrolyzed urethane compounds and demonstrated esterase and protease activities (Liu et al. 2021). Photoinitiation processes are essential for the biodegradation of PUR in the environment, just like they are for polymers with a C–C backbone. Thh_Est from T. halotolerans and the thermoactive Est119 from T. alba were categorized as esterases (Hu et al. 2010; Ribitsch et al. 2012). Prior research suggested that lipases or cutinases were more effective at hydrolyzing the polyester surface than esterases were at breaking down polyesters (Ribitsch et al. 2012). It is hypothesized that esterase activity cleaves PUR. The urethanase bacterial enzyme hydrolyzed urethane compounds and demonstrated esterase and protease activities (Liu et al. 2021). Photoinitiation processes are essential for the biodegradation of PUR in the environment, just like they are for polymers with a C–C backbone. Hydroperoxides are formed due to this oxidation, which takes place in the α-methylene position (Gewert et al. 2015). The ester or urethane linkages can then be hydrolyzed by other enzymatic processes, with the latter being catalyzed at slower breakdown rates. It is known that aryl esters or carbamates are necessary for the catalytic activity of urethanes. However, it is unclear how these enzymes recognize their substrates (Chen et al. 2020). Abiotic factors are not strictly necessary for good biodegradation because PET and some purr polymers can be biodegraded into specified oligomers and monomers thanks to hydrolyzable ester or amide linkages (Wei et al. 2020). Hydroperoxides are formed due to this oxidation, which takes place in the α-methylene position (Gewert et al. 2015). The ester or urethane linkages can then be hydrolyzed by other enzymatic processes, with the latter being catalyzed at slower breakdown rates. It is known that aryl esters or carbamates are necessary for the catalytic activity of urethanes. However, it is unclear how these enzymes recognize their substrates (Chen et al. 2020). Abiotic factors are not strictly necessary for good biodegradation because PET and some purr polymers can be biodegraded into specified oligomers and monomers thanks to hydrolyzable ester or amide linkages (Wei et al. 2020).

2.4   Recent Advances in Plastic Degradation by Enzymes

Enzymes recently studied in enzymatic degradation of plastics comprise mainly esterase, cutinase, lipase, and PETases (Kyrikou and Briassoulis 2007). All these enzymes are involved in degrading different plastic polymers discussed above

(Yamashita et al. 2019). Esterase enzyme from the yeast Pseudozyma Antarctica has been recently shown to accelerate the degradation process of plastic mulch films, which were kept in laboratory conditions. Mulch films were made from biodegradable plastic, which showed a high degradation rate when the strain was grown in the presence of xylose (Closas et al. 2016). The plastic films were incubated with the enzyme for 12 weeks, and the SEM analysis showed that the smooth surface of the movie had holes pierced through them, resulting in an uneven surface.

Furthermore, according to Barragan et al. (2016), the amorphous portion of the films deteriorated more quickly than the crystalline portion; according to Thirunavukarasua et al. (2008), a lipase enzyme that was isolated from the yeast Cryptococcus sp. MTCC 5455, which was cultured on leftover agricultural waste, showed a highly respectable potential in the hydrolysis of polybutylene succinate (PBS) and polybutylene succinate-co-adipate (PBSA). Sixty-two hours for PBS and sixteen hours for PBSA showed a total breakdown of the plastic components (Thirunavukarasu et al. 2016). To determine which chemical groups were broken down by the lipase enzyme, Fourier Transform Infrared Spectroscopy (FTIR) was applied to the PBS and PBSA polymers that were exposed to enzymatic degradation (Kijchavengkul et al. 2010). When compared to the PBS film that was not damaged, a notable drop in the transmittance of the C–O and -C–O–C– bonds was seen. A prominent peak at 3372 cm1, which relates to the stretching vibration of the –OH group, indicated the major chain scission at ester linkage (Umare et al. 2007). The stretching vibrations of the C–H and O–C–O bonds, respectively, were represented by these peaks. The absence of the 1, 4-butanediol peak during enzymatic degradation suggested that the carboxyl end of the polymer chain may have hydrolyzed, according to the NMR study of the PBS and PBSA polymer films (Li et al. 2011). A different lipase derived from Candida rugosa was employed to break down the poly (butylene succinate-co-hexamethylene succinate) copolymer. According to the research, a copolymer's sensitivity to a lipase enzyme attack increases with its HS content. Weight loss was a metric to track the P (BS-co-HS) copolymer's breakdown (Pereira et al. 2001). According to Hu et al. (2016), Fusarium solani's cutinase has also been overexpressed in Pichia pastoris, which breaks down PBS plastic. The PBS films were fully broken down only 6 h after the recombinant cutinase was able to speed up the breakdown process (Marija and Jasna 2001). The PBS films' emergence of holes verified that the enzyme's maximum effect was achieved by lengthening its breakdown duration.

The recombinant enzyme broke down both the crystalline and amorphous structures of PBS, according to the analysis of the PBS films subjected to enzymatic attack (Maeda et al. 2005). It was hypothesized in a previous study on the degradation of PBS films with varying molecular weights that the films with larger molecular weights could degrade more quickly and readily (Pan et al. 2018). Numerous different biodegradable plastic kinds have been demonstrated to be broken down by cutinases (Masaki et al. 2005). High molecular weight plastic material, based on polylactic acid (PLA), was broken down by an enzyme that functions similarly to cutinase and isolated from yeast that belonged to Cryptococcus sp. Strain S-2 (Van Gemeren et al. 1998). Compared to other known enzymes that also break down PLA plastic, this enzyme broke down the material in 60 h at a low concentration of 0.8 μg/ml of enzyme (Oda et al. 2000). The hydrolase class of enzymes includes the esterases, which include lipase and cutinase. Lately, it has been shown that combining these two enzymes will work well for the biological breakdown of polycaprolactone (PCL), a plastic (Liu et al. 2019). After using an end-to-end fusion approach to generate the fusion of these enzymes, Pichia pastoris was used to overexpress the fusion system (Aris et al. 2016). Following a 6-h duration, the PCL degradation facilitated by the Lip-Cut fusion enzyme surpassed the individual degradation rates of these two enzymes (Khan et al. 2017). The two primary byproducts of PCL degradation were 6-hydroxyhexanoic acid and a tiny ratio of \-caprolactone, according to analysis of the degraded products using the GC-MS technique (Shah et al. 2013). It has been studied whether a PHB depolymerise may break down polyhydroxybutyrate (PHB), polybutylene succinate (PBS), and polyethylene succinate (PES) (Jung et al. 2018). Aspergillus fumigatus was used to extract this enzyme, which was then cloned and purified (Kasuya et al. 1998). This enzyme is an extracellular enzyme operating on PHB, PBS, and PES. It functions best at a pH of 6.5 and a temperature of 55 degrees Celsius. The strains that displayed a clear zone on agar plates with PHB, PBS, or PES as the only carbon source were used to isolate this enzyme (Hoang et al. 2007). The strain was utilized to extract the depolymerase enzyme, as evidenced by the clear zone, which also showed that it could break down different types of plastic (Tokiwa and Calabia 2004). When the plastic films were incubated with depolymerase for 17 h, a weight loss of almost 95% was seen (Ohtaki et al. 2006). The films' severe deterioration was indicated by a shift in surface topology from smooth to fibrous, with potholes measuring 3–5 mm (Dickinson 1998). The numerous recent enzymatic degradation systems covered above are summarized in Table 2.1. Before Joo et al. (2018), it was believed that polyethylene terephthalate (PET) plastic, utilized to make the majority of throwaway plastic items, was not biodegradable. A Japanese scientific team has discovered a bacterium that thrives on pet plastic garbage that has been building up for a long time (Webb et al. 2013). It was determined that Ideonella sakaiensis was the bacterial strain that could increase on PET waste bottles by using PET as the primary carbon source for its growth. PETase, an enzyme, allowed the microorganism to break down PET plastic into useful monomers of ethylene glycol and terephthalic acid. Through hydrophobic interactions, the enzyme's active site was able to readily accept this artificial polymer, which allowed it to trigger the hydrolytic destruction of PET plastic (Paci and La Mantia 1999). Cutinase enzymes isolated from Humicola insolens (HiC) and the thermostable Thermobifida cellulosilytica cutinase 1 (Cut) were used to carry out the enzymatic degradation of PBTF (poly (1,4-butylene 2,5-thiophene-dicarboxylate) and PBF (poly (1,4-butylene 2,5-furan dicarboxylate) films. Enzymatic hydrolyses were performed in 1 M K_2HPO_4/KH_2PO_4 buffer for approximately 72 h at a five μM enzyme concentration. It has been noted that maintaining the pH at eight yields the greatest results for film deterioration. The effectiveness of the enzyme activity was demonstrated by the demonstration of a 20% weight loss in both films. PBF degraded a little bit more quickly than PBTF. HiC and Cut, the two studied enzymes, showed greater activity at higher

Table 2.1 Enzymes used for plastic degradation in recent years

Microorganism	Enzyme extracted	Method of detection	Plastic material degraded	Reference
Ideonella sakaiensis	PETase and MHETase	Examination of degradation product	PET	Knott et al. (2020)
Tritirachium album L	Lipase/proteinase K	Percentage weight loss and examination of degradation product (lactic acid)	Cellulose nanocrystal composites	Hegyesi et al. (2019)
Thermobifida cellulosilytica	Cutinase	Percentage weight loss and scanning electron microscopy	PBTF and PBF	Giglia et al. (2019)
Pseudozyma Antarctica	Esterase	Scanning electron microscopy	Biodegradable plastic	Yamashita et al. (2019)
Streptomyces sp. SM14	Esterase	Plate clearing assay	Synthetic polyester-degrading activity	Almeida et al. (2019)
Cryptococcus sp	Lipase	FTIR, NMR	PBS	Thirunavukarasu et al., (2016)
Ideonella sakaiensis 201-F6	Aromatic polyesterase	Scanning electron microscopy	Poly(ethylene terephthalate) degradation	Yoshida et al. (2016)
Rhodococcus sp. TMP2	Monoxygenases	Reverse transcription polymerase chain reaction	Polyethylene degradation	Takei et al. (2008)
Alcaligenes faecalis	Lipase	High-performance liquid chromatography	Polycaprolactone degradation	Oda et al. (1997)
Sphingomonas sp. strain 113P3	PVA dehydrogenase	Reverse transcription polymerase chain reaction	Polyvinyl alcohol degradation	Hirota-Mamoto et al. (2006)
Bacillus subtilis	Esterase	SDS-PAGE	Polyurethane degradation	Rowe and Howard (2002)

temperatures (65 C vs. 50 C), and HiC's hydrolysis kinetics were faster than Cut's (Giglia et al. 2019). In a different study, four distinct enzymes—Protease K, Protease, Esterase, and Lipase—were examined on the breakdown of PLA reinforced with flax fiber. Scanning electron microscopy (SEM), morphological investigations of PLA reinforced with flax fiber revealed that the enzyme Proteinase K had the greatest impact on the degradability of the polymers. The PLA reinforced with 30% flax fiber showed a mass loss of about 52%, while the plain PLA polymer showed just a 21% mass loss. Thus, flax fiber improved the polymer's biodegradability. The biggest alteration to the PLA polymer's general structure was also brought about by proteinase K. According to Stepczynska and Rytlewski (2018), protease and lipase enzymes had the least effect on the breakdown of polymers.

Candida rugosa's lipase enzyme and Tritirachium album's proteinase K enzyme degraded PLA polymer and its nanocomposites reinforced with cellulose nanocrystals (CNC). The results for the enzyme proteinase K were encouraging, but the study indicates that the enzyme lipase was ineffective in catalyzing the breakdown of this polymer. The reaction mixture's pH dropped due to the lactic acid that formed throughout the reaction, which led to the enzymes' breakdown. It was discovered that a lower ionic strength was more advantageous for the polymer's breakdown. As the sample polymer broke down in 3 days, the cellulose nanocrystals that corroborated the PLA polymer improved the breakdown rate (Hegyesi et al. 2019). PBS/PLA blend films enhanced with benzyl peroxide were also degraded using the enzyme Proteinase K. Before deterioration studies, the mix films had a smooth surface visible with scanning electron microscopy, or SEM. Little cracks started appearing on the film's surface after the enzyme was treated for 12 h, and the roughness continued to rise for another 24 h (Hu et al. 2018). For the conversion of amorphous PET film to monomeric units (terephthalic acid and ethylene glycol), a highly synergistic interaction between PETase enzyme and MHETase enzyme from I. sakaiensis was reported (Knott et al. 2020). In the presence of these two enzymes at varying doses, the degree of hydrolysis for PET at 30 C was monitored for 96 h. When PET film was broken down, MHETase showed no activity by itself. The product (terephthalic acid and ethylene glycol) releases scales with PETase loading upon the addition of MHETase to the reaction with PETase enzyme, but at a significantly higher level than with PETase alone. As the concentration of both enzymes grew, more component monomers were liberated, suggesting that these reactions were enzyme-limited instead of substrate-limited. The synergy investigation showed that even at low concentrations compared to PETase, the presence of MHETase supports the degradation scales with the PETase enzyme.

2.5   Environmental Implications of Enzymatic Degradation of Synthetic Plastics

Because plastics come from a variety of origins and travel through a variety of freshwater and terrestrial ecosystems, they are ubiquitous in a variety of settings. The ultimate destiny of plastic garbage is the result of various processes: (i) The buildup of plastic materials in environmental compartments is hazardous to human health and some ecosystems. This is particularly true for the most widely used polymers, which frequently add extra chemical agents, additives, and solubilizers to enhance their mechanical and physical properties (Danso et al. 2018). (ii) Abiotic oxidation can happen in various ways, depending on the temperature, oxygen content, and UV light (Mohanan et al. 2020). This outcome suggests that plastic polymers have partially degraded or activated and that hazardous substances, such as volatile organic compounds (VOCs), have also been released and may interact with the surrounding ecology (La Nasa et al. 2021). (iii) Although most plastic trash has always been disposed of in landfills, chemical or mechanical recycling and inciner-ation are the conventional treatment techniques for plastic garbage when correctly collected and transferred to waste treatment plants (Zhang et al. 2021). While the waste-to-energy method has also been used to recover energy from waste treatment, it only handles a small portion of polymer waste. It produces ash loaded with heavy metals and hazardous gases, including furans and dioxins, among other adverse environmental effects. (iv) The most environmentally benign techniques that still need to be shown are recycling and biodegradation technologies that use microor-ganisms or their enzymes (Lee and Liew 2021). Microbial communities may adapt and use metabolic transformations to break down polymers into simpler chemicals in settings contaminated by plastic. The microbial attack involves multiple processes: Biodeterioration damages the polymer's surface and is accelerated by environmental and abiotic variables. Depolymerization occurs during the fragmentation of the biodeterioration plastics, producing oxidized or lower molecular weight plastic pieces. Microorganisms that grow biofilms on plastic surfaces and generate active catalytic enzymes capable of breaking down or biotransforming synthetic polymers trigger the assimilation pathways. Last but not least, the oxidized plastic derivatives are carried into the cells during the mineralization process, where various enzymatic reactions result in total breakdown into oxidized metabolites, including CO_2 and H_2O (Amobonye et al. 2021). Therefore, plastic polymer biodegradation would be favored in certain environmental conditions and when microbial populations have the enzymatic arsenal to handle them.

2.6 Conclusion and Future Aspects

Enzymes are highly capable of breaking down many kinds of plastic materials, both biodegradable and non-biodegradable, that contaminate the environment and endan-ger all living things on Earth. To this day, numerous researchers are working to develop novel and enhanced processes that can break down different types of plastic in the presence of enzymes isolated from diverse microbiological sources. Most research has found that hydrolytic enzymes mainly cause the degradation of plastic polymer. Recent studies in this field have shown that increasing the activity of the enzymes that have been shown to break down different kinds of plastic is all that is now required to turn enzymatic plastic degradation into a commercial process. One strategy to improve the stability of plastic-degrading enzymes and enable them to function in harsh environmental circumstances, such as extremely hot or cold temperatures, is to manipulate the active site of the enzymes. One technique that can be used to modify the structure and functional groups of the enzyme's active site so that it can more readily accept plastic polymer and function in various pH and temperature ranges is site-directed mutagenesis.

Additionally, the range of plastic degradability in a single experiment can be expanded, and more than two types of plastics can be accommodated by combining two enzymes in a combined bifunctional system. This was found when PCL plastic material was broken down using a bifunctional Lipase-Cutinase enzyme system. It is imperative to standardize and enhance the efficiency of the microbial species isolation procedure that bears these plastic-degrading enzymes. This will enable the screening of pure breeding colonies of these microorganisms in the shortest amount of time. By increasing the amount of enzyme produced by pure colonies, standardizing this first step can shorten the time it takes to break down the plastic polymer.

References

Alessandra P, Cinzia P, Paola G et al (2010) Heterologous laccase production and its role in industrial applications. Bioeng Bugs 1:252–262. https://doi.org/10.4161/bbug.1.4.11438

Almeida EL, Carrillo Rincón AF, Jackson SA, Dobson ADW (2019) *In silico* screening and heterologous expression of a polyethylene terephthalate hydrolase (PETase)-like enzyme (SM14est) with Polycaprolactone (PCL)-degrading activity, from the marine sponge-derived strain *Streptomyces* sp. SM14. Front Microbiol 10:2187. https://doi.org/10.3389/fmicb.2019.02187

Al-Tammar KA, Omar O, Abdul Murad AM, Abu Bakar FD (2016) Expression and characterization of a cutinase (AnCUT2) from Aspergillus niger. Open Life Sci 11:2938. https://doi.org/10.1515/biol-2016-0004

Amobonye A, Bhagwat P, Singh P, Pillai S (2021) Plastic biodegradation: frontline microbes and their enzymes. Sci Total Environ 759:143536. https://doi.org/10.1016/j.scitotenv.2020.143536

Andler R, Valdés C, Díaz-Barrera A, Steinbüchel A (2020) Biotransformation of poly(cis-1,4-isoprene) in a multiphase enzymatic reactor for continuous extraction of oligo-isoprenoid molecules. New Biotechnol 58:10–16. https://doi.org/10.1016/j.nbt.2020.05.001

Aris MH, Annuar MSM, Ling TC (2016) Lipase-mediated degradation of polyε-caprolactone in toluene: behavior and its action mechanism. Polym Degrad Stab 133:182–191

Barragan DH, Pelacho AM, Closas LM (2016) Degradation of agricultural biodegradable plastics in the soil under laboratory conditions. Soil Res 54:216–224

Carniel A, Valoni E, Nicomedes J, da Conceição Gomes A, Castro AM (2016) Lipase from Candida Antarctica (CALB) and cutinase from Humicola insolens act synergistically for PET hydrolysis to terephthalic acid. Process Biochem 59:84–90

Carniel A, Valoni É, Nicomedes J et al (2017) Lipase from Candida antarctica (CALB) and cutinase from Humicola insolens act synergistically for PET hydrolysis to terephthalic acid. Process Biochem 59:84–90. https://doi.org/10.1016/j.procbio.2016.07.023

Chen CC, Han X, Ko TP et al (2018) Structural studies reveal the molecular mechanism of PETase. FEBS J 285:3717–3723. https://doi.org/10.1111/febs.14612

Chen CC, Dai L, Ma L, Guo RT (2020) Enzymatic degradation of plant biomass and synthetic polymers. Nat Rev Chem 4:114–116. https://doi.org/10.1038/s41570-020-0163-6

Closas LM, Costa J, Cirujeda A, Aibar J, Zaragoza C, Pardo A, Suso ML, Moreno MM, Moreno C, Lahoz I, Macua JI, Pelacho AM (2016) Above-soil and in-soil degradation of oxo- and bio-degradable mulches: a qualitative approach. Soil Res 54:225–236

Cowan AR, Costanzo CM, Benham R et al (2022) Fungal bioremediation of polyethylene: challenges and perspectives. J Appl Microbiol 132:78–89. https://doi.org/10.1111/jam.15203

Danso D, Schmeisser C, Chow J et al (2018) New insights into the function and global distribution of polyethylene terephthalate (PET)-degrading bacteria and enzymes in marine and terrestrial metagenomes. Appl Environ Microbiol 53:1689–1699. https://doi.org/10.1128/AEM.02773-17

Dickinson E (1998) Proteins at interfaces and in emulsions stability, rheology and interactions. J Chem Soc Faraday Trans 94:1657–1669

Gewert B, Plassmann MM, Macleod M (2015) Pathways for degradation of plastic polymers floating in the marine environment. Environ Sci Process Impacts 17:15131521. https://doi.org/10.1039/C5EM00207A

Giglia M, Quartinellob F, Soccioc M, Pellisb A, Lottic N, Georg MG, Licocciaa S, Munari A (2019) Enzymatic hydrolysis of poly(1,4-butylene 2,5-thiophenedicarboxylate) (PBTF) and poly (1,4-butylene 2,5-furandicarboxylate) (PBF) films: a comparison of mechanisms. Environ Int 130:104852

Gomez PV, Sarah E, Fabres CJ (2018) Marine plastic pollution as a planetary boundary threat-the drifting piece in the sustainability puzzle. Mar Pol 96:213–220

Gricajeva A, Nadda AK, Gudiukaite R (2022) Insights into polyester plastic biodegradationby carboxyl ester hydrolases. J Chem Technol Biotechnol 97:359–380. https://doi.org/10.1002/jctb.6745

Gross RA, Kalra B (2002) Biodegradable polymers for the environment. Science 297:803–807

Hegyesi N, Zhang Y, Kohari A, Polyaka P, Sui X, Pukanszky B (2019) Enzymatic degradation of PLA/cellulose nanocrystal composites. Ind Crop Prod 141:111799

Herrero Acero E, Ribitsch D, Dellacher A et al (2013) Surface engineering of a cutinase from Thermobifida cellulosilytica for improved polyester hydrolysis. Biotechnol Bioeng 110:2581–2590. https://doi.org/10.1002/bit.24930

Hirota-Mamoto R, Nagai R, Tachibana S, Yasuda M, Tani A, Kimbara K, Kawai F (2006) Cloning and expression of the gene for periplasmic poly (vinyl alcohol) dehydrogenase from Sphingomonas sp. strain 113P3, a novel-type quinohaemoprotein alcohol dehydrogenase. Microbiology 152(7):1941–1949

Hoang KC, Tseng M, Shu WJ (2007) Degradation of polyethylene succinate (PES) by a new thermophilicMicrobispora strain. Biodegradation 18(3):333–342

Hu X, Thumarat U, Zhang X et al (2010) Diversity of polyester-degrading bacteria in compost and molecular analysis of a thermoactive esterase from Thermobifida alba AHK119. Appl Microbiol Biotechnol 87:771–779. https://doi.org/10.1007/s00253-010-2555-x

Hu X, Gao Z, Wang Z, Su T, Yang L, Li P (2016) Enzymatic degradation of poly(butylene succinate) by cutinase cloned from Fusarium solani. Polym Degrad Stab 134:211–219

Hu X, Su T, Li P, Wang Z (2018) Blending modification of PBS/PLA and its enzymatic degradation. Polym Bull 75:533–546

Hung CS, Zingarelli S, Nadeau LJ et al (2016) Carbon catabolite repression and impranil polyurethane degradation in Pseudomonas protegens strain Pf-5. Appl Environ Microbiol 82:6080–6090. https://doi.org/10.1128/AEM.01448-16

Janusz G, Pawlik A, Swiderska-Burek U et al (2020) Laccase properties, physiological functions, and evolution. Int J Mol Sci 21:966. https://doi.org/10.3390/ijms21030966

Joo S, Cho J, Seo H, Francis Son H, Sagong H-Y, Shin TJ, Choi SY, Lee SY, Kim KJ (2018) Structural insight into molecular mechanism of polyethylene terephthalate degradation. https://doi.org/10.1038/s41467-018-02881-1

Jung HW, Yang MK, Su R-C (2018) Purification, characterization, and gene cloning of an Aspergillus fumigates polyhydroxy butyrate depolymerase used for degradation of polyhydroxybutyrate, polyethylene succinate, and polybutylene succinate. Polym Degrad Stab 154:186–194

Kamboj M (2016) Degradation of plastics for clean environment. Int J Adv Res Eng Appl Sci 5(3):10–19

Kasuya KI, Takagi K-i, Ishiwatari SI, Yoshida Y (1998) Biodegradabilities of various aliphatic polyesters in natural waters. Polym Degrad Stab 59:327–332

Kawabata T, Oda M, Kawai F (2017) Mutational analysis of cutinase-like enzyme, Cut190, based on the 3D docking structure with model compounds of polyethylene terephthalate. J Biosci Bioeng 124:28–35. https://doi.org/10.1016/j.jbiosc.2017.02.007

Khan I, Dutta JR, Ganesan R (2017) Lactobacillus sps. lipase mediated poly (epsiloncaprolactone) degradation. Int J Biol Macromol 95:126–131

Kijchavengkul T, Auras R, Rubino M, Selke S, Ngouajio M, Fernandez RT (2010) Biodegradation and hydrolysis rate of aliphatic aromatic polyester. Polym Degrad Stab 95:2641–2647

Knott BC, Erickson E, Allen MD, Gado JE, Graham R, Kearns FL, Pardo I, Topuzlu E, Anderson JJ, Austin HP, Dominick G, Johnson CW, Rorrer NA, Szostkiewicz CJ, Copie V, Payne CM, Lee Woodcock H, Donohoe BS, Beckham GT, McGeehan JE (2020) Characterization and engineering of a twoenzyme system for plastics depolymerisation, 25476–25485. Proc Natl Acad Sci USA 117:41

Kyrikou I, Briassoulis D (2007) Biodegradation of agricultural plastic films: a critical review. J Polym Environ 15:125–150

La Nasa J, Lomonaco T, Manco E, Ceccarini A, Fuoco R, Corti A, Modugno F, Castevetro V, Degano I (2021) Plastic breeze: volatile organic compounds (VOCs) emitted by degrading macro- and microplastics analyzed by selected ion flow-tube mass spectrometry. Chemosphere 270:128612. https://doi.org/10.1016/j.chemosphere.2020.128612

Lecourt M, Chietera G, Bernard Blerot B, Antoniotti S (2021) Laccase-catalyzed oxidation of allylbenzene derivatives: towards a green equivalent of ozonolysis. Molecules 26:6053. https://doi.org/10.3390/molecules26196053

Lee A, Liew MS (2021) Tertiary recycling of plastics waste: an analysis of feedstock, chemical and biological degradation methods. J Mater Cycles Waste Manag 23:32–43. https://doi.org/10.1007/s10163-020-01106-2

Li F, Hu X, Guo Z, Wang Z, Wang Y, Liu D, Xia H, Chen S (2011) Purification and characterization of a novel poly(butylene succinate)-degrading enzyme from Aspergillus sp. XH0501-a. World J Microbiol Biotechnol 27:2591–2596

Liu B, He L, Wang L et al (2018) Protein crystallography and site-direct mutagenesis analysis of the poly(ethylene terephthalate) hydrolase petase from Ideonella sakaiensis. Chembiochem 19: 1471–1475. https://doi.org/10.1002/cbic.201800097

Liu M, Zhang T, Long L, Zhang R, Ding S (2019) Efficient enzymatic degradation of poly (ε-caprolactone) by an engineered bifunctional lipase-cutinase. Polym Degrad Stab 160:120–125

Liu J, He J, Xue R et al (2021) Biodegradation and up-cycling of polyurethanes: progress, challenges, and prospects. Biotechnol Adv 48:107730. https://doi.org/10.1016/j.biotechadv.2021.107730

Maeda H, Yamagata Y, Abe K, Hasegawa F, Machida MR, Gomi K et al (2005) Purification and characterization of a biodegradable plastic-degrading enzyme from Aspergillus oryzae. Appl Microbiol Biotechnol 67:778–788

Marija SN, Jasna D (2001) Synthesis and characterization of biodegradable poly (− butylene succinate-co-butylene adipate). Polym Degrad Stab 74:263–270

Masaki K, Kamini NR, Ikeda H, Iefuji H (2005) Cutinase-like enzyme from the yeast Cryptococcus sp. strain S-2 HydrolyzesPolylactic acid and other biodegradable plastics. Appl Environ Microbiol 71:7548–7550

Mohanan N, Montazer Z, Sharma PK, Levin DB (2020) Microbial and enzymatic degradation of synthetic plastics. Front Microbiol 11:580709. https://doi.org/10.3389/fmicb.2020.580709

Nakkabi A, Sadik M, Fahim M, Ittobane N, Koraichi SL, Barkai H, El abed, S. (2015) Biodegradation of polyester urethanes by Bacillus subtilis. Int J Environ Res 9(1):157–162

Oda Y, Oida N, Urakami T, Tonomura K (1997) Polycaprolactone depolymerase produced by the bacterium Alcaligenes faecalis. FEMS Microbiol Lett 152(2):339–343

Oda Y, Yonetsu A, Urakami T, Tonomura K (2000) Degradation of polylactide by commercial proteases. J Polym Environ 8:29–32

Ohtaki S, Maeda H, Takahashi T, Yamagata Y, Hasegawa F, Gomi K, Nakajima T, Abe K (2006) Novel hydrophobic surface binding protein, HsbA, produced by Aspergillus oryzae. Appl Environ Microbiol 72(4):2407–2413

Paci M, La Mantia FP (1999) Influence of small amounts of polyvinylchloride on the recycling of polyethylene terephthalate. Polym Degrad Stab 63:11–14

Pan W, Bai Z, TingtingSu WZ (2018) Enzymatic degradation of poly(butylenesuccinate) with different molecular weights by cutinase. Int J Biol Macromol 111:1040–1046

Pavia DL, Lampman GM, Kriz GS (1988) Introduction to organic laboratory techniques, 3rd edn. Saunders, Fort Worth

Pereira EB, De Castro HF, De Moraes FF, Zanin GM (2001) Kinetic studies of lipase from Candida rugosa. Appl Biochem Biotechnol 91:739

Pometto AL, Lee BT, Johnson KE (1992) Production of an extracellular polyethylenedegrading enzyme(s) by Streptomyces species. Appl Environ Microbiol 58:731–733

Popov KV, Knyazev VD (2014) Molecular dynamics simulation of C-C bond scission in polyethylene and linear alkanes: effects of the condensed phase. J Phys Chem A 118:2187–2195. https://doi.org/10.1021/jp411474u

Restrepo-Flórez JM, Bassi A, Thompson MR (2014) Microbial degradation and deterioration of polyethylene – a review. Int Biodeterior Biodegrad 88:83–90. https://doi.org/10.1016/j.ibiod.2013.12.014

Ribitsch D, Acero EH, Greimel K et al (2012) A new esterase from Thermobifida halotolerans hydrolyses polyethylene terephthalate (PET) and polylactic acid (PLA). Polymers 4:617–629. https://doi.org/10.3390/polym4010617

Ribitsch D, Hromic A, Zitzenbacher S, Zartl B, Gamerith C, Pellis A, Jungbauer A, Łyskowski A, Steinkellner G, Gruber K, Tscheliessnig R, Herrero Acero E, Guebitz GM (2017) Small cause, large effect: structural characterization of cutinases from Thermobifida cellulosilytica. Biotechnol Bioeng 114(11):2481–2488. https://doi.org/10.1002/bit.26372

Ronkvist ÅM, Xie W, Lu W, Gross RA (2009) Cutinase-catalyzed hydrolysis of poly(ethylene terephthalate). Macromolecules 42:5128–5138. https://doi.org/10.1021/ma9005318

Rowe L, Howard GT (2002) Growth of bacillus subtilis on polyurethane and the purification and characterization of a polyurethanase-lipase enzyme. Int Biodeterior Biodegrad 50:33–40

Saravanan A, Kumar PS, Vo D-VN et al (2021) A review on catalytic-enzyme degradation of toxic environmental pollutants: microbial enzymes. J Hazard Mater 419:126451. https://doi.org/10.1016/j.jhazmat.2021.126451

Seo JS, Keum YS, Li QX (2009) Bacterial degradation of aromatic compounds. Int J Environ Res 6:278–279

Shah AA, Hasan F, Hameed A, Ahmed S (2008) Biological degradation of plastics: a comprehensive review. Biotechnol Adv 26:246–265

Shah AA, Eguchi T, Mayumi D, Kato S, Shintani N, Kamini NR, NakajimaKambe T (2013) Degradation of aliphatic and aliphaticearomatic co-polyesters by depolymerases from Roseateles depolymerans strain TB-87 and analysis of degradation products by LC-MS. Polym Degrad Stab 98:2722–2729

Shimao M (2001) Biodegradation of plastics. Curr Opin Biotechnol 12:242–247

Singh B, Sharma N (2008) Mechanistic implications of plastic degradation. Polym Degrad Stab 93:561–584. https://doi.org/10.1016/j.polymdegradstab.2007.11.008

Stepczynska M, Rytlewski P (2018) Enzymatic degradation of flax-fibers reinforced polylactide. Int Biodeterior Biodegrad 126:160–166

Takei D, Washio K, Morikawa M (2008) Identification of alkane hydroxylase genes in Rhodococcus sp. strain TMP2 that degrades a branched alkane. Biotechnol Lett 30(8):1447–1452. https://doi.org/10.1007/s10529-008-9710-9

Tang KHD, Lock SSS, Yap P-S et al (2021) Immobilized enzyme/microorganism complexes for degradation of microplastics: a review of recent advances, feasibility and future prospects. Sci Total Environ 832:154868. https://doi.org/10.1016/j.scitotenv.2022.154868

Thirunavukarasu K, Purushothaman S, Sridevi J, Aarthy M, Gowthaman MK, Kambe TN, Kamini NR (2016) Degradation of poly(butylene succinate) and poly(butylene succinate-co-butylene adipate) by a lipase from yeast Cryptococcus sp. grown on agro-industrial residues. Int Biodeterior Biodegrad 110:99–107

Thirunavukarasua K, Edwinolivera NG, Durai Anbarasana S, Gowthamana MK, Iefujib H, Kamini NR (2008) Removal of triglyceride soil from fabrics by a novel lipase from Cryptococcus sp. S-2. Process Biochem 43:701–706

Tokiwa Y, Calabia BP (2004) Degradation of microbial polyesters. Biotechnol Lett 26:1181–1189

Trevino AL, Gutiérrez-Sánchez G, Rodríguez-Herrera R, Aguilar CN (2012) Microbial enzymes involved in polyurethane biodegradation: a review. J Polym Environ 20:258–265

Tsushima S, Matsushita Y (2010) Technical report on the PCR-DGGE analysis of bacterial and fungal soil communities. National Institute for Agro-Environmental Sciences, Tsukuba. ver. 3.3

Twala PP, Mitema A, Baburam C, Feto NA (2020) Breakthroughs in the discovery and use of different peroxidase isoforms of microbial origin. AIMS Microbiol 6:330349. https://doi.org/10.3934/microbiol.2020020

Umare SS, Chandure AS, Pandey RA (2007) Synthesis, characterization and biodegradable studies of 1, 3-propanediol based polyesters. Polym Degrad Stabil 92:464–479

Van Gemeren IA, Beijersbergen A, Van den Hondel CAMJJ, Verrips CT (1998) Expression and secretion of defined cutinase variants by Aspergillus awamori. Appl Environ Microbiol 64: 2794–2799

Villain F, Coudane J, Vert M (1995) Thermal degradation of polyethylene terephthalate: study of polymer stabilization. Polym Degrad Stab 49:393–397

Vona IA, Costanza JR, Cantor HA, Roberts WJ (1965) Manufacture of plastics, vol 1. Wiley, New York, pp 141–142

Waring RH, Harris RM, Mitchell SC (2018) Plastic contamination of the food chain: a threat to human health? Maturitas 115:64–68

Webb HK, Arnott J, Crawford RJ, Ivanova EP (2013) Plastic degradation and its environmental implications with particular reference to poly(ethylene terephthalate). Polymers 5:1–18

Wei R, Zimmermann W (2017) Microbial enzymes for the recycling of recalcitrant petroleum-based plastics: how far are we? Microb Biotechnol 10:1308–1322. https://doi.org/10.1111/1751-7915.12710

Wei R, Oeser T, Zimmermann W (2014) Synthetic polyesterhydrolyzing enzymes from thermophilic actinomycetes. Adv Appl Microbiol 89:267–305. https://doi.org/10.1016/B978-0-12-800259-9.00007-X

Wei R, Tiso T, Bertling J et al (2020) Possibilities and limitations of biotechnological plastic degradation and recycling. Nat Catal 3:867–871. https://doi.org/10.1038/s41929-020-00521-w

Wilkes RA, Aristilde L (2017) Degradation and metabolism of synthetic plastics and associated products by Pseudomonas sp.: capabilities and challenges. J Appl Microbiol 123:582–593. https://doi.org/10.1111/jam.13472

Williams DF (1981) Enzymic hydrolysis of polylactic acid. Eng Med 10:5–7

Xu J, Cui Z, Nie K et al (2019) A quantum mechanism study of the C-C bond cleavage to predict the bio-catalytic polyethylene degradation. Front Microbiol 12:489. https://doi.org/10.3389/fmicb.2019.00489

Yamashita YS, Ueda H, Koitabashi M, Kitamoto H (2019) Pretreatment with an esterase from the yeast Pseudozyma Antarctica accelerates biodegradation of plastic mulch film in soil under laboratory conditions. J Biosci Bioeng 127(1):93–98

Yoshida S, Hiraga K, Takehana T, Taniguchi I, Yamaji H, Maeda Y, Toyohara K, Miyamoto K, Kimura Y, Oda K (2016) A bacterium that degrades and assimilates poly(ethylene terephthalate). Science 351:1196–1199

Zhang F, Zhao Y, Wang D et al (2021) Current technologies for plastic waste treatment: a review. J Clean Prod 282:124523

Chapter 3
The Potential of Enzyme Engineering to Positively Impact Environmental Sustainability

Abstract Sustainable catalytic technology development and use are essential to achieve our net-zero ambitions. Here, we summarize how modified enzymes—directed evolution being a particular emphasis—can enhance the sustainability of various processes and contribute to environmental preservation. Carbon dioxide (CO_2) can now be captured by robust and efficient biocatalysts integrated into novel, effective metabolic pathways for CO_2 fixation. Enzymes have been improved for Bioremediation, increasing their capacity to break down hazardous substances. With modified cutinases and PETases created for the depolymerization of the ubiquitous plastic, polyethylene terephthalate (PET), biocatalytic recycling is gathering momentum. Lastly, using optimized enzymes to transform plant biomass into biofuels or other high-value products is a growing biocatalytic technique for gaining access to petroleum-based feedstocks and chemicals. We intend to demonstrate through these examples how biocatalysis and enzyme engineering can help build a more efficient and clean chemical industry.

Keywords Enzyme engineering · Bioremediation · Enzyme immobilization · Metagenomics · Proteomics

3.1 Introduction

Over 8 billion people live on Earth today; by 2050, that number is expected to increase to 9.7 billion. Significant questions about how to advance environmental sustainability while yet meeting the requirements of the populace are raised by this increase. We are facing an ecological crisis due to the extraordinary demands that human activity has placed on our planet. The energy, chemicals, and materials necessary for contemporary life are produced mainly through fossil fuels, which has resulted in an unsustainable increase in greenhouse gas emissions. A concerning rise in deforestation and significant biodiversity losses are caused by the proliferation of carbon-intensive farming practices brought on by the increased demand for food. Health-harming toxic substances are far too frequently released into the environment due to accidental spills or as standard operating procedures.

J. A. Parray et al., *Enzymes in Environmental Management*, SpringerBriefs in Environmental Science, https://doi.org/10.1007/978-3-031-74874-5_3

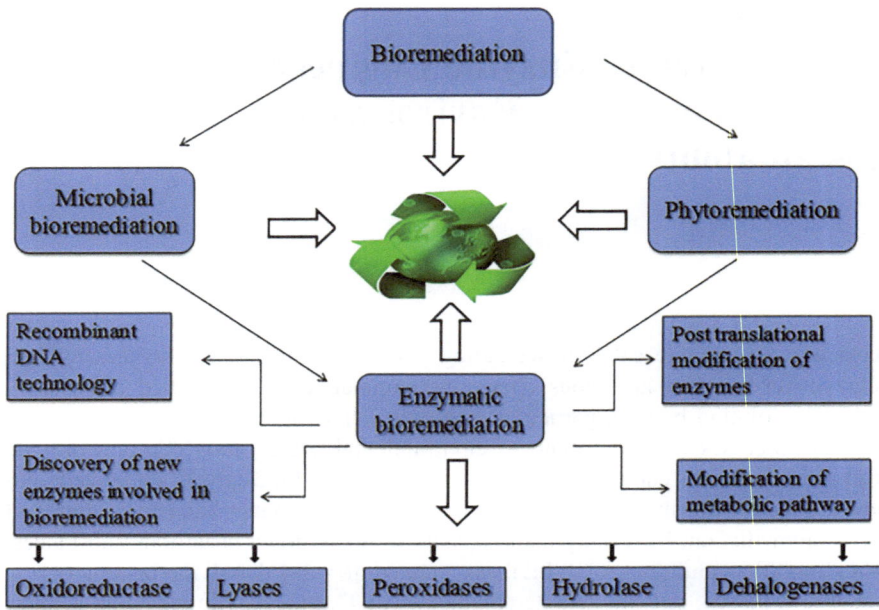

Fig. 3.1 An overview of combinatory methodologies for bioremediation

Moreover, it is frequently possible to integrate several enzymes into "one-pot" multi-step cascades that enable complex chemical transformations with significantly better step economy, resulting in higher productivity and lower solvent and energy input requirements. Enzymes are widely used because of one crucial characteristic: they are very engineerable. We can now construct biocatalytic systems without being constrained by natural enzymes because of advancements in protein engineering techniques. Alternatively, enzymes can now be specially engineered to fulfill the requirements of intended uses. In this study, we will first provide a brief overview of enzyme engineering, emphasizing directed evolution. Then, we will highlight significant instances where enzymes have been created to tackle environmental problems. Instead of offering a thorough overview of every activity, the article aims to demonstrate how enzyme engineering can positively impact environmental sustainability through a few chosen instances. Different fields of Bioremediation, such as microbial and enzymatic, including phytoremediation, and their strategies are depicted in Fig. 3.1.

3.2   Enzyme Engineering and Directed Evolution

The features of wild-type enzymes must first be optimized, as they are frequently unsuitable for the intended use (Chen and Arnold 2020). Enzymes can be modified to improve selectivity and kinetic parameters, increase substrate specificity or range, increase tolerance to immobilization, and improve stability in process-relevant conditions like high substrate loadings, high temperatures, or organic cosolvents.

For enzyme engineering, directed evolution is a flexible and popular approach that simulates Darwinian evolution on lab scales and circumstances. This method allows for the simultaneous optimization of several enzyme parameters, even if there is a need for a thorough understanding of the structure and mechanism of the enzyme. To travel sequence space more quickly and lessen the screening burden, directed evolution is occasionally employed in conjunction with computational tools like machine learning (ML) and protein sequence activity relationships (ProSAR) (Wu et al. 2019). According to Lovelock et al. (2022), computational techniques help enhance protein stability, rationally re-engineering substrate binding sites or even introducing novel catalytic activities that can be further improved through evolution. Iterative cycles of gene expression, enzyme library member screening, and DNA library production are all part of the directed evolution cycle. Once an appropriate starting template has been found, DNA libraries are created using standard molecular biology methods like site saturation or random mutagenesis. By introducing DNA libraries into cells, it is possible to separate library members spatially and make a connection between phenotype and genotype. Maintaining this relationship is necessary to enable the characterization of individual library members during the manufacturing and screening of proteins. Single colonies are frequently arranged into multi-well plates to produce and screen proteins. Here, they can be assessed by various chromatographic, spectrophotometric, and spectroscopic methods. Sometimes, throughput can be increased, and a more thorough investigation of protein sequence space is possible by employing more specialized screening and selection-based techniques. Nevertheless, it is not feasible to link the necessary enzyme activity to either a fluorescence output or cell viability in many situations using such methods. After the library is evaluated, the best-performing variants are separated, sequenced, and used as templates for further rounds of evolution.

3.3   Enzyme Engineering and Bioremediation

Microorganisms' long process of pollutant degradation makes Bioremediation less practical in real-world settings. To get around the restrictions above, microbial enzymes isolated from their cells have been utilized in Bioremediation in recent years rather than employing entire microorganisms (Thatoi et al. 2014). According to Kalogerakis et al. (2017), enzymes are intricate biological macromolecules that catalyze a variety of biochemical events that are part of the pathways leading to the breakdown of pollutants. By reducing the activation energy of molecules, enzymes can speed up a reaction. Enzymes can be classified into the following kinds according to their capacity for biodegradation, as shown in Table 3.1. Bioremediation based on purified and partially purified enzymes does not depend on the growth of a particular microorganism in a polluted environment, but it depends upon the catalytic activity of the enzyme secreted by microbes. Bioremediation can be

Table 3.1 Enzymes involved in Bioremediation and their functions

Enzyme classification	Examples	Functions	Reference
Oxidoreductase	Monooxygenase	Incorporation of an oxygen atom into the substrate and utilize substrate as a reducing agent. Desulfurization, dehalogenation, denitrification, ammonification, and hydroxylation of substrate	Karigar and Rao (2011)
	Dioxygenase	The introduction of two oxygen atoms to the substrate results in estradiol cleaving and estradiol cleaving with the formation of an aliphatic product	Karigar and Rao (2011)
	Laccases	Cleave rings in aromatic compounds reduce one molecule of oxygen in the water and produce free radicals.	Shraddha et al. (2011)
	Peroxidases	Catalyze reduction reaction in the presence of peroxides, such as hydrogen peroxide (H_2O_2), and generate reactive free radicals after oxidation of organic compounds.	Bansal and Kanwar (2013)
Hydrolases	Lipases	Break triglycerol into glycerol and fatty acid, which are widely used for wastewater treatment, polyaromatic hydrocarbon degradation, etc.	Mehta et al. (2021)
	Cellulases	Break down complex cellulosic materials into simple sugars, commonly used in treating agricultural residues such as cotton waste, sawdust of Khaya ivorensis, and rice.	Bhardwaj et al. (2017)
	Carboxylesterases	Catalyzes the hydrolysis of carboxyl ester bonds present in synthetic pesticides such as organophosphates with the addition of water	Singh (2014)
	Phosphotriesterases	Catalyze hydrolysis of phosphodiester, the main components of organophosphorus compounds used worldwide in pesticides, causes severe poisoning and death.	Santillan et al. (2016)
	Haloalkane dehalogenase	Used for biodegradation of halogenated aliphatic compounds such as 1,2,3-trichloro propane	Nagata et al. (2015)
	Protease	Enzymes that hydrolyze peptide bonds in aqueous environment	Karigar and Rao (2011)

possible in soil with poor nutrients using a purified enzyme. Toxic side products produced by microbial biotransformation are not created using enzymatic biotransformation, which is safe for the environment. Compared to microbes, enzymes are more specific to their substrate and mobile in nature because of their smaller size. To explain this in brief, different properties of an enzyme that can be modified by enzyme engineering are described in Fig. 3.2.

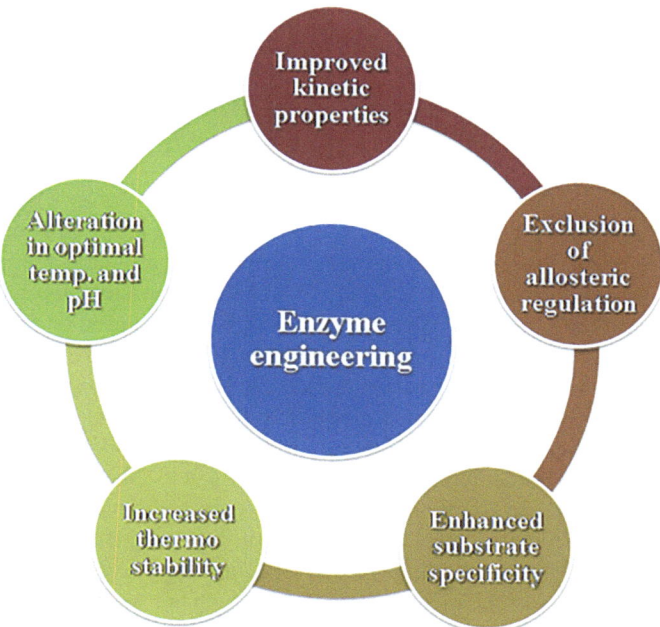

Fig. 3.2 Enzyme engineering to improve enzymatic properties for bioremediation

3.4   Oxidoreductases

Different types of bacteria, fungi, and higher plants generate and secrete oxidoreductases, which are used in oxidative coupling—a process that involves oxidizing molecules by transferring electrons from reductants to oxidants—and release methanol, CO_2, and chloride ions as a result. Microorganisms use the heat or energy the oxidoreductases produce to break down contaminants for their metabolic processes. Oxidoreductases have been employed to degrade numerous contaminants, both natural and manufactured. Bacillus safensis CFA-06, a gram-positive bacteria, creates oxidoreductase to break down petroleum molecules. In nature, lignin breakdown results in various phenolic compounds that are then transformed into other forms by oxidoreductases via polymerization and co-polymerization through bonding with other substances (Husain 2006). Textile industry-produced color compounds are released into the environment, broken down by several oxidoreductases, including laccase and peroxidases (Novotny et al. 2004). According to reports, white rot fungus Panus tigrinus and its extracellular oxidoreductase, laccase, Independent peroxidase, and lignin peroxidase eliminated phenols, color, and organic load from olive-mill effluent. Redox processes in numerous bacterial species result in the release of oxidoreductase enzymes, which are necessary for the reduction of radioactive metals. The plant, a member of the Fabaceae, Gramineae, and Solanaceae families, secrete extracellular enzymes that break down pollutants in the soil, like hydrocarbons comprising petroleum and chlorinated chemicals (Newman et al. 1998).

3.5   Oxygenases

The primary enzymes involved in the aerobic breakdown of aromatic compounds are called oxygenases; they catalyze the ring cleavage in aromatic compounds by introducing one or two oxygen molecules, which is a necessary step in the degradation process. Oxygenases are divided into two subclasses based on the number of oxygen molecules they catalyze: monooxygenases, which catalyze the addition of a single oxygen atom, and dioxygenases, which catalyze the addition of two oxygen atoms. Pesticide bioremediation uses glyphosate oxidase (GOX), isolated from the bacterium Pseudomonas sp. LBr. According to Scott et al. (2008), GOX transforms glyphosate into aminomethylphosphonate (AMPA) and liberates the keto acid glyoxylate. By adding an oxygen molecule, oxygenases can improve the reactivity and solubility of a wide range of substrates, including chlorinated compounds. FAD, NADH, and NADPH compounds are cofactors for oxygenase action. Herbicides, fungicides, and pesticide chemicals containing halogen are widely distributed in the environment and are broken down by oxygenase through the catalysis of their dehalogenation process. Additionally, some marine bacteria generate the enzyme oxygenase, which aids in the breakdown of organic contaminants.

3.6   Monooxygenases

By introducing a single oxygen molecule, monooxygenases help to break down aromatic chemicals and increase their solubility and reactivity. Monooxygenases have reportedly been linked to the hydroxylation, desulphurization, denitrification, and dehalogenation of aromatic substances (Arora et al. 2010). Based on the cofactor they employ, monooxygenases are divided into two groups: P450 monooxygenases and flavin-dependent monooxygenases. The two members of the TC-FDM (two-component flavin diffusible monooxygenase) family, Esd (endosulfan diol) and Ese (endosulfan ether), are responsible for the degradation of pesticides containing chlorine, including endosulfan (Bajaj et al. 2010). A tightly bound flavin cofactor is present in the Flavin-dependent monooxygenase family, reduced by NAD (P)H. Both prokaryotes and eukaryotes include the heme-containing enzyme P450 monooxygenase. P450 monooxygenase, isolated from the Bacillus megaterium BM3 bacteria, can break down many substrates, including aromatic chemicals and fatty acids (Roccatano 2015). Methane monooxygenase metabolizes halides that contain heavy metals, aromatic chemicals, and aliphatic molecules. There are two types of methane monooxygenase: one in the cytoplasm and the other in the cytoplasmic membrane. According to Singh and Singh (2017), soluble MMOs derived from the bacteria Methylocella palustris can break down various contaminants, including hydrocarbon, aliphatic, and aromatic chemicals. Certain monooxygenases, such as quinol monooxygenase (YgiN) from Escherichia coli and tetracenomycin F1 monooxygenase (TcmH) from Streptomyces glauscens, have also been identified and isolated; these enzymes are cofactor-free (Arora et al. 2010). Another significant member of the monooxygenase family that is employed in numerous sectors to oxidize pollutants produced is cytochrome P450. In both prokaryotes and eukaryotes, P450 oxidoreductase has more than 200 subfamilies. The fungus P. Chrysosporium, which causes white rot, has been fully sequenced using genome sequencing. The results show that the fungus has 150 genes organized into 16 gene clusters, categorized under 12 CYP families, 11 fungal CYP clans, and one P450 reductase component. Every P450 oxidoreductase member has an iron-containing porphyrin group, and they all need a non-covalently bound cofactor—most commonly, NAD(P)H—to recycle their redox center. Another kind of P450 oxidoreductase that is present in the liver of mammals and aids in the breakdown of herbicides such as atrazine, norflurazon, and chlortoluron is called cytochrome CYP1A1 (Kawahigashi et al. 2005). To catalyze the oxidative dealkylation of phenyl urea herbicides like linuron, cytochrome CYP76B1, which was isolated from Helianthus tuberosus (Jerusalem artichoke), was cloned and expressed in tobacco and Arabidopsis. Using the ice-nucleation protein from Pseudomonas syringae, researchers have also created a whole-cell biocatalyst that expresses rat NADPHecytochrome P450 oxidoreductase on the surface of E. coli. This biocatalyst has a variety of uses, such as the selective synthesis of novel chemicals and pharmaceuticals, bioconversion, Bioremediation, and the creation of biochips (Yim et al. 2006).

3.7 Dioxygenases

Dioxygenases catalyze the breakdown of aromatic pollutants by introducing two oxygen molecules into the ring. Arachidonic ring hydroxylation dioxygenases (ARHDs) and aromatic ring cleavage dioxygenases (ARCDs) are the categories into which aromatic dioxygenases can be divided based on their function. While ARCDs cleave compounds' aromatic rings, ARHDs break down chemical compounds by adding two molecules of oxygen into the ring (Parales and Ju 2011). Pseudomonas putida F1 produces toluene dioxygenase (TOD), which catalyzes the breakdown of toluene. This multi-component enzyme sees either monooxygenase or dioxygenase activity. It functions as a dioxygenase for many different types of pollutants, including aromatics and aliphatic chemicals. According to Mukherjee and Roy (2013), TOD is a monooxygenase for olefins with an aliphatic chain, monocyclic aromatic compounds, and other molecules. The catechol dioxygenases cause the biotransformation of aromatic precursors into aliphatic products present in soil bacteria. According to Ali et al. (2017), ring-opening 2,4 dioxygenases aid in the Bioremediation of quinaldine, and 1H-4-isoquinoline catalyzes the breakage of two carbon-carbon bonds that result in the production of carbon monoxide. The chemical, dye, and pharmaceutical industries discharge numerous aromatic chemicals into the environment. The aromatic ring at the 1, 2-position is broken down by dioxygenase to allow two oxygen molecules to enter the substrate. For instance, Pseudomonas putida's naphthalene dioxygenase is involved in the breakdown of naphthalene.

3.8　Laccases

Laccases are oxidases containing copper and catalyze the oxidation of various aromatic and phenolic chemicals in soil and water. It is present in several bacterial, fungal, insect, and plant isoforms. Although they can be secreted extracellularly, laccases are continuously synthesized in the cell and have the ability to break down lignin, aryl groups containing diamines, amino groups containing phenols, and ortho and para diphenols. Laccase oxidizes its bonds and causes azo dyes to lose color and become less hazardous environmental chemicals (Legerska et al. 2016). Two laccase isozymes isolated from the fungus Trametes hispida were the cause of the decolorization activity. Strong and Claus (2011) state that R. practicola produces laccase, which can biotransform and destroy phenolic chemicals. According to Dodor et al. (2004), laccase immobilization on solid support lengthens its half-life, stability, and resistance to protease enzymes. Laccase isolated from the fungus Trametes versicolor was immobilized on porous glass beads. It is shown to be a potent enzyme for the Bioremediation of several contaminants, including aromatic heterocyclic compounds, phenolic compounds, and aromatic compounds containing amines. By removing electrons from the organic substrate, laccase can lower the dioxygen molecules of pollutants in the water. Synthetic laccase was also employed in the paper and pulp industries to improve the bleaching of pulp and textiles. For docking experiments, the two-dimensional structures of contaminants retrieved from the NCBI database were compared to the X-ray crystal structures of laccases deposited in the Protein Data Bank (PDB). The two-dimensional structures of the pollutants were transformed into three-dimensional structures using an online program called CORINA. Moreover, protein-ligand docking was done using GOLD. The best average GOLD fitness score for the bacterial and fungal laccase enzymes, respectively, was found in about 30% and 17% of the datasets that were chosen, indicating that laccase may be able to oxidize these pollutants.

3.9　Peroxidases

Peroxidases are ubiquitous and are produced by animals, plants, fungi, and bacteria. Peroxidases help degrade lignin and other aromatic contaminants by using hydrogen peroxide as a mediator. Phenolic radicals are produced by the oxidation of phenolic compounds and aggregates and become less soluble and precipitate quickly. Heme can be present in the enzyme (Bansal and Kanwar 2013). The heme-containing peroxidases can be divided into two groups: one group is found only in animals, and the other group is found in fungi, bacteria, and plants. Peroxidases found in bacteria, fungi, and plants are further divided into three classes: intracellular enzyme found in class 1, including cytochrome c peroxidase produced by yeast, ascorbate peroxidase (APX) made by some species of plants and bacterial catalase peroxidases. Fungal enzymes that are secreted, such as manganese peroxidase (MnP) and lignin peroxidase (LiP), are classified as class 2. Horseradish peroxidases (HRP), secreted by horseradish plants, are examples of plant-secreted peroxides found in class 3. Non-heme peroxidases include manganese catalase, haloperoxidase, thiol peroxidase, alkylhydroperoxidase, and NADH peroxidase, belonging to five distinct families (Koua et al. 2009). Of all the enzymes studied, lignin peroxidase and manganese peroxidase show the most potential for breaking down harmful compounds. Enzyme aggregates immobilized by horseradish peroxidases and cross-linked by the use of ethylene glycol-bis succinic acid N-hydroxy succinimide (EG-NHS) as a cross-linking agent were created. A packed bed reactor system was used to test the biodegradation efficiency of HRP-CLEAs. Moreover, 2,4,6-trichlorophenol, a carcinogen, and hazardous pollutant undergoes oxidative para-dechlorination due to HRP. The breakdown of thiazole compounds was investigated using soybean peroxidase and chloroperoxidase.

3.10   Lignin Peroxidases

Bacteria and fungi like Phanerochaete versicolor and Phanerochaete chrysosporium release lignin peroxidases (LiPs), which are monomeric heme-containing proteins and secondary metabolites. Fe (III) is pentacoordinated in LiPs with four heme tetrapyrrole nitrogens and a histidine residue. In the presence of mediator Ike veratryl alcohol and cosubstrate hydrogen peroxide, it catalyzes the oxidation of hazardous pollutants. In the first phase of LiPs, the natural ferric enzyme [Fe (III)] was oxidized by H_2O_2 twice, producing compound I. Compound II was then reduced by a reducing substrate after compound I gained an electron. When the enzyme returns to its original ferric oxidation state in the final phase, compound II takes up a second electron from the reduced substrate. LiPs have a lot of uses in the bioremediation and wastewater treatment fields.

Regarding specificity and thermostability, bacterial peroxidases are more effective at breaking down lignin than fungal peroxidases. LiPs typically break down lignin, a component of plant cell walls. LiP enzymes can oxidize aromatic compounds with redox potentials more significant than 1.4 V (Piontek et al. 2001).

3.11   Manganese Peroxidases

Manganese peroxidases (MnPs) are extracellular heme-containing enzymes that break down lignin and can oxidize Mn2þ into Mn3þ through several steps. Enzyme MnP has several acidic amino acid residues and a single heme group containing a manganese binding site. Mn2þ also functions as the best-reducing substrate, adding one electron to compound I of MnP. It is thought that this chelator acts indirectly to break down xenobiotics and lignin. According to Have and Teunissen (2001), they catalyze the breakdown of certain phenols and aromatic compounds, including amines and dyes. Trametes sp. 48424, a white rot fungus, was used to identify and purify MnPTra-48424. The MnP-Tra-48424 enzyme can decolorize many dyes, including indigo, anthraquinone, azo, and triphenylmethane. When paired with heavy metal ions and organic solvents, it can also degrade other colors, including methyl green, indigo carmine, and methyl green. MnP-Tra48424 purified also degrades several polycyclic aromatic hydrocarbons (PAHs) (Zhanga et al. 2016). P. incarnata KUC8836 was identified to harbor the gene (pimp1) encoding manganese-dependent peroxidase during the breakdown of anthracene. To improve the bioremediation process, this gene was further expressed in Saccharomyces cerevisiae fungi (Lee et al. 2016). Additionally, MnP was immobilized using chitosan beads activated by glutaraldehyde, which suggests a higher potential for decolorizing textile sector dye effluent. Immobilization on a sol-gel matrix made by the fungus Ganoderma lucidium IBL 05 improved the stability and half-life of MnP.

3.12   Hydrolases

The most popular application of hydrolytic enzymes is the Bioremediation of herbicides and insecticides, which lowers their toxicity. Hydrolytic enzymes break central chemical bonds, including peptide bonds, carbon-halide bonds, and esters. Bioremediation is a safer and more cost-effective method for the breakdown of harmful organic compounds than physico-chemical therapy (Karigar and Rao 2011). The Bioremediation of organic polymers—toxic substances with molecular weights less than 600 Da that can fit through cell pores—is catalyzed by an extracellular hydrolase released by bacteria. Hydrolytic enzymes, oil spills, organophosphates, and carbamate pesticides can be remediated successfully. The food and chemical industries use extracellular hydrolytic enzymes such as proteases, lipases, xylanases, DNA, and amylase. Biomass breakdown is accomplished by the enzymes glycosidase, cellulase, and hemicellulase. To boost the amounts of the enzyme, Lei et al. (2017) cloned the hydrolyzing enzyme gene from Microbacterium sp. del-6F into Escherichia coli BL21 (DE3) using carbendazim, a commonly used fungicide. This enzyme could hydrolyze Carbamazim to produce 2-aminobenzimidazole. In a different investigation, the gram-negative bacteria's outer membrane vesicles were attached to the organophosphorus hydrolase enzyme, which has substantial activity in breaking down parathion chemical compounds (Su et al. 2017).

3.13   Lipases

Natural lipases catalyze the conversion of triacylglycerols into glycerol and free fatty acids, essential components of hydrocarbons. Lipases are found throughout nature. Numerous types of bacteria, plants, actinomycetes, and mammalian cells produce lipases. Lipases are involved in hydrolysis, inter-esterification, esterification, alcoholics, and aminolysis processes. Lipase activity reduced the amount of hydrocarbon in the polluted soil. According to Ghafil et al. (2016), these enzymes hydrolyze fatty acids into triglycerol, diacylglycerol, monoacylglycerol, and glycerol. The media was optimized using several statistical techniques to increase the microbial lipase output. The medium used to produce a new lipase from the fungus Pseudomonas aeruginosa SL-72 that breaks down crude oil was optimized using statistical techniques to facilitate the Bioremediation of crude oils.

3.14   Cellulases

The most prevalent biopolymer on Earth, cellulose, is broken down by essential enzymes called cellulases. Microorganisms can produce extracellular, cell-bound, and cell envelope-associated cellulase (Yang et al. 2016). In the industries that create detergents, cellulose microfibrils generated during operations are eliminated using cellulases. Certain Bacillus strains produce alkaline cellulases, while Humicola and Trichoderma fungi produce neutral and acidic cellulases (Hmad and Gargouri 2017). These cellulases have been used in the paper and pulp industries for the Bioremediation of ink during paper recycling (Karigar and Rao 2011). Because they are more suited for demanding industrial processes in a variety of industries, such as food, textile, laundry, pulp and paper, biofuel production, and waste management, scientists are particularly interested in the alkaline cellulases produced by marine microorganisms (Annamalai et al. 2016). The Humicola species produces cellulase, which is highly adapted to extreme environmental factors, including high pH and temperature. This enzyme is employed in the detergent and washing powder industries to break down hydrogen bonds (Imran et al. 2016). According to Aslam et al. (2017), the bacterium B. amyloliquefacience-ASK11 was isolated from an industrial waste containing leather tanning and produced 20 U mL1 of cellulases at 10 ppm of Cr (VI).

3.15   Carboxylesterases

The enzyme carboxylesterases have catalyzed the breakdown of natural and artificial products, including organophosphates, carbamate ester bonds, and chlorine compounds. In the fields, carboxylesterases have been utilized to break down herbicides, insecticides, and fungicide sprays. On the outer membrane of E. coli, carboxylesterases E2 from strain Pseudomonas aeruginosa PA1 were discovered to be involved in mercury absorption at the polluted site (Yin et al. 2016). For the past 40 years, synthetic pyrethroids—a significant family of insecticides—have been utilized in agricultural settings. Carboxylesterases have dissolved their ester bond using a single mechanism, breaking down all varieties of pyrethroid pesticides. Heidari et al. (2005) reported that the active site of carboxylesterases, isolated from Lucilia cuprina and Drosophila melanogaster, was modified through in vitro mutagenesis to facilitate the hydrolysis of pyrethroids. The degradation of organophosphorous pesticides is carried out by carboxylesterases E3, which were identified by L. cuprina (Scott et al. 2008).

3.16   Phosphotriesterases

Pesticides like parathion used in agricultural areas and chemical waste emitted from industry can be degraded by phosphotriesterases. Herbicides and insecticides contain the chemical parathion, which includes an organophosphate. Phosphotriesterases, sometimes called aryldialkylphosphatase and organophosphorus hydrolase, break down organophosphate, an ester of phosphoric acid. Restaino et al. (2016) developed and purified three recombinant thermostable phosphotriesterase: SacPox, which was obtained from Sulfolobus acidocaldarius, and SsoPox W263F and C258L/I261F/W263A, whose gene was recovered from wild type Sulfolobus solfataricus. Certain marine bacterial strains, including Ruegeria mobilis, Thalassospira tepidiphila, and Phaeobacter sp., can break down the phosphate triester seen in coastal ocean settings. The Geobacillus stearothermophilus (GsP) produced a new enzyme homologous to phosphotriesterases and capable of hydrolyzing compounds that include lactone and organophosphates. The bacteria Geobacillus stearothermophilus (GsP) produced phosphotriesterase-like lactonase (PLL), which is exceptionally thermostable and active at 100 C (Hawwa et al. 2009).

3.17   Enzyme Engineering and Sustainable Plastic Recycling

3.17.1   Mining Plastic-Degrading Enzymes via Metagenomics-Based Approaches

The field of metagenomics has shown great promise in aiding in the identification of novel enzymes from diverse ecological settings. Most known enzymes that break down plastic have been found using the traditional culture-dependent method. The culture-dependent approach entails the following steps: strain taxonomy classification, identification of putative enzymes using molecular biological or computational methods, and enrichment and isolation of microorganisms expressing the desired enzyme under suitable cultivation conditions (Kawai et al. 2014). However, because less than 1% of all organisms on the Earth are thought to have been grown, the culture-dependent method severely restricts the potential to discover new plastic-degrading enzymes. On the other hand, the metagenomic technique that does not rely on culture has become an effective means of investigating the great majority of microorganisms found in various environmental settings. Numerous genes encoding enzymes that may depolymerize various plastic materials have been extracted from many ambient metagenome samples. Thus, using metagenomic approaches, we have recently made headway in identifying new enzymes that break down plastic from the vast pool of naturally occurring biocatalysts.

Choosing the right screening techniques is one of these crucial aspects of metagenomic mining. Sequence-based and function-based screening are generally employed for the metagenomic library. By looking through bioinformatic databases, sequence-based screening uses functional gene annotation and sequence similarity comparison. For instance, a search strategy driven by a hidden Markov model was used to find a poly(ethylene terephthalate) (PET) hydrolytic enzyme (PET2) by in silico sequence-based screening from metagenome databases. More recently, metagenomes collected from landfills have yielded several gene sequences that resemble those encoding recognized enzymes that can break down polyurethane (PU) plastics. In silico, sequence-based metagenomic screening is a somewhat quick and economical method for enzyme mining. However, the quantity and caliber of gene annotation in the databases of known plastic-degrading enzymes that are currently available restricts its potential. Additionally, novel families of plastic-degrading enzymes with minimal sequence similarity to previously identified ones may go unnoticed by this method. Furthermore, additional characterization and validation of the functioning of the enzymes is required because sequence similarities do not ensure plastic-degrading activity. Function-based screening searches metagenomic libraries for the targeted traits using activity assays as an alternative. Compared to sequence-based screening, this method is especially beneficial for mining entirely new classes of enzymes whose sequences differ more from those of homologous ones already in existence. For instance, function-based agar plate assays were used to screen several enzymes phylogenetically belonging to completely new esterase families from environmental metagenomes. These enzymes demonstrated hydrolytic activity towards various polyesters, such as poly(lactic acid) (PLA), poly(ε-caprolactone) (PCL), and poly(butylene succinate-co-adipate) (PBSA). The capacity of conventional agar plate assays to screen sizable metagenomic libraries is limited. New bacteria and enzymes that break down plastic may be found more quickly because of recent research on high-throughput screening techniques (Weinberger et al. 2012).

3.18   Conclusion and Future Prospective

A viable path toward a more sustainable future is the development of biocatalytic techniques powered by modified enzymes for process design, replacement, or supplementation. The review's examples demonstrate how directed evolution might yield optimized biocatalysts that can drastically improve many processes. Certain designed enzymes, like those for carbon fixation, plastic recycling, and lignocellulose deconstruction, have already achieved or are approaching economic application. Even with these achievements, many obstacles must be addressed before biocatalysis and enzyme engineering become extensively utilized for sustainable purposes. For example, there are several processes for which protein engineering is challenging, or enzymes are not readily available. The creation and application of biocatalytic recycling techniques for substitute plastics like nylons and Bioremediation techniques to remove additional hazardous and environmentally resistant compounds like poly-fluoroalkyl substances (PFAS) are important examples. Furthermore, many examples have yet to be carried out in large-scale industrial settings, and it is unclear how biocatalytic processes might be applied in the framework of currently available infrastructure. Lastly, there are situations where enzymatic methods could not be economically viable or provide the desired sustainability benefits.

New enzyme design, discovery, and optimization approaches—including deep learning-based approaches—hold enormous promise for swiftly broadening the spectrum of chemistries that may be accessed to address these issues. Technologically advanced ultrahigh throughput screening assays will expedite the engineering and discovery of new enzymes, and novel analyte identification techniques will help identify complex analytes quickly. Pilot or industrial-scale process experiments are necessary to increase reactions and enhance feasibility. These tests should be combined with further rounds of protein optimization to address any shortcomings. We can further improve process efficiency and sustainability by investigating the integration of biocatalysis with new or existing complementary chemical techniques. This will enable us to maximize the advantages of various technologies. Lastly, life cycle and techno-economic evaluations should be used to confirm the sustainability benefits and financial consequences of replacing existing processes with biocatalytic alternatives. The results of these evaluations are also essential for defining process operating settings that are both environmentally friendly and financially viable, as these circumstances serve as target parameters for enzyme engineering. The world's environmental issues must be addressed quickly, and humanity needs to develop new strategies and concepts. We are confident that biocatalysis can be a significant factor in resolving these issues and assisting us in achieving our sustainability objectives while preserving the environment.

References

Ali MI, Khatoon N, Jamal A (2017) Polymeric pollutant biodegradation through microbial oxido-reductase; a better strategy to safe environment. Int J Biol Macromol 105:9–16. https://doi.org/10.1016/j.ijbiomac.2017.06.047

Annamalai N, Rajeswari MV, Balasubramanian T (2016) Thermostable and alkaline cellulases from marine sources. In: New and future developments in microbial biotechnology and bioengineering, p 91e98. https://doi.org/10.1016/B978-0-444-63507-5.00009-5. ISBN 9780444635075

Arora PK, Srivastava A, Singh VP (2010) Application of monooxygenases in dehalogenation, desulphurization, denitrification and hydroxylation of aromatic compounds. J Bioremed Biodegr 1:112. https://doi.org/10.4172/2155-6199.1000112

Aslam S, Hussain A, Qazi JI (2017) Production of cellulase by Bacillus amyloliquefaciens-ASK11 under high chromium stress. Waste Biomass Valoriz 1e9:53–61

Bajaj A, Pathak A, Mudiam MR, Mayilraj S, Manickam N (2010) Isolation and characterization of a Pseudomonas sp. strain IITR01 capable of degrading endosulfan and endosulfan sulfate. J Appl Microbiol 109(6):2135e2143

Bansal N, Kanwar SS (2013) Peroxidase(s) in environment protection. Sci World J. https://doi.org/10.1155/2013/714639

Bhardwaj V, Degrassi G, Bhardwaj RK (2017) Bioconversion of cellulosic materials by the action of microbial cellulases. Int Res J Eng Technol 4(8):494–503

Chen K, Arnold FH (2020) Engineering new catalytic activities in enzymes. Nat Catal 3:203–213

Dodor DE, Hwang HM, Ekunwe SI (2004) Oxidation of anthracene and benzo[a] pyrene by immobilized laccase from Trametes versicolors. Enzyme Microbial Technol 35:3210e3217. https://doi.org/10.1016/j.enzmictec.2004.04.007

Ghafil JA, Hassan SS, Zgair AK (2016) Use of immobilized lipase in cleaning up soil contaminated with oil. World J Exp Biosci 4:53e57

Have RT, Teunissen PJ (2001) Oxidative mechanisms involved in lignin degradation by white-rot fungi. Chem Rev 101:3397e3413

Hawwa R, Aikens J, Turner RJ, Santarsiero BD, Mesecar AD (2009) Structural basis for thermostability revealed through the identification and characterization of a highly thermostable phosphotriesterase-like lactonase from Geobacillus stearothermophilus. Arch Biochem Biophys 488:109e120. https://doi.org/10.1016/j.abb.2009.06.005

Heidari R, Devonshire AL, Campbell BE, Dorrian SJ, Oakeshott JG, Russell RJ (2005) Hydrolysis of pyrethroids by carboxylesterases from Lucilia cuprina and Drosophila melanogaster with active sites modified by in vitro mutagenesis. Insect Biochem Mol Biol 35(6):597–609

Hmad BI, Gargouri A (2017) Neutral and alkaline cellulases: production, engineering, and applications. J Basic Microbiol 57(8):653–658

Husain Q (2006) Potential applications of the oxidoreductive enzymes in the decolorization and detoxification of textile. Crit Rev Biotechnol 26:201e221. https://doi.org/10.1080/07388550600969936

Imran M, Anwar Z, Irshad M, Asad MJ, Ashfaq H (2016) Cellulase production from species of fungi and bacteria from agricultural wastes and its utilization in industry: a review. Adv Enzym Res 4(02):44–55

Kalogerakis N, Fava F, Corvine PF (2017) Bioremediation advances. N Biotech 38:41e42. https://doi.org/10.1016/j.nbt.2017.07.004

Karigar CS, Rao SS (2011) Role of microbial enzymes in the bioremediation of pollutants: a review. Enzym Res 7:1–11. https://doi.org/10.4061/2011/805187

Kawahigashi H, Hirose S, Ohkawa H, Ohkawa Y (2005) Transgenic rice plants expressing human CYP1A1 remediate the triazine herbicides atrazine and simazine. J Agric Food Chem 53:8557e8564. https://doi.org/10.1021/jf051370f

Kawai F et al (2014) A novel Ca2+-activated, thermostabilized polyesterase capable of hydrolyzing polyethylene terephthalate from Saccharomonospora viridis AHK190. Appl Microbiol Biotechnol 98:10053–10064

Koua D, Cerutti L, Falquet L et al (2009) PeroxiBase: a database with new tools for peroxidase family classification. Nucleic Acids Res 37:D261eD266

Lee AH et al (2016) Heterologous expression of a new manganese-dependent peroxidase gene from Peniophora incarnata KUC8836 and its ability to remove anthracene in Saccharomyces cerevisiae. J Biosci Bioeng 122:716e721. https://doi.org/10.1016/j.jbiosc.2016.06.006

Legerska B, Chmelova D, Ondrejovic M (2016) Degradation of synthetic dyes by laccase – a mini review. Nova Biotechnol Chim 15:90. https://doi.org/10.1515/nbec-2016-0010

Lei J, Wei S, Ren L, Hu S, Chen P (2017) Hydrolysis mechanism of carbendazim hydrolase from the strain Microbacterium sp. djl-6F. J Environ Sci 54:171e177–171e177. https://doi.org/10.1016/j.jes.2016.05.027

Lovelock SL, Crawshaw R, Basler S, Levy C, Baker D, Hilvert D, Green AP (2022) The road to fully programmable protein catalysis. Nature 606:49–58

Mehta A, Guleria S, Sharma R, Gupta R (2021) The lipases and their applications with emphasis on food industry. In: Microbial biotechnology in food and health. Academic Press, pp 143–164

Mukherjee P, Roy P (2013) Cloning, sequencing and expression of novel trichloroethylene degradation genes from Stenotrophomonas maltophilia PM102: a case of gene duplication. J Bioremed Biodegr 4:2. https://doi.org/10.4172/2155-6199.1000177

Nagata Y, Ohtsubo Y, Tsuda M (2015) Properties and biotechnological applications of natural and engineered haloalkane dehalogenases. Appl Microbiol Biotechnol 99(23):9865e9881

Newman LA, Doty SL, Gery KL, Heilman PE, Muiznieks I, Shang TQ, Siemieniec ST, Strand SE, Wang X, Wilson AM, Gordon MP (1998) Phytoremediation of organic contaminants: a review of phytoremediation research at the University of Washington. J Soil Contam 7(4):531e542. https://doi.org/. https://doi.org/10.1080/10588339891334366

Novotny C, Svobodova K, Erbanova P, Cajthamal T, Kasinath A, Lang E, Sasek V (2004) Ligninolytic fungi in bioremediation: extracellular enzyme production and degradation rate. Soil Biol Biochem 36:1545e1551

Parales RE, Ju KS (2011) Degradation of aromatic hydrocarbons. In: Comprehensive biotechnology, 2nd edn, p 115e134

Piontek K, Smith AT, Blodig W (2001) Lignin peroxidase structure and function. Biochem Soc Trans 29:111e116

Restaino OF, Borzacchiello MG, Ilaria S, Fedele L, Schiraldi C (2016) Biotechnological process design for the production and purification of three recombinant thermophilic phosphotriesterases. N Biotech 33:15. https://doi.org/10.1016/j.nbt.2016.06.776

Roccatano D (2015) Structure, dynamics, and function of the monooxygenase P450 BM-3: insights from computer simulations studies. J Phys Condens Matter 27:273102. https://doi.org/10.1088/0953-8984/27/27/273102

Santillan JY, Dettorre LA, Lewkowicz ES, Iribarren AM (2016) New and highly active microbial phosphotriesterase sources. Microbiol Lett 363(24):fnw276

Scott C, Pandey G, Hartley CJ, Jackson CJ, Cheesman MJ, Taylor MC, Pandey R, Khurana JL, Teese M, Coppin CW, Weir KM (2008) The enzymatic basis for pesticide bioremediation. Indian J Microbiol 48(1):65–79

Singh B (2014) Review on microbial carboxylesterase: general properties and role in organophosphate pesticides degradation. Biochem Mol Biol 2:1–6

Singh JS, Singh DP (2017) Methanotrophs: an emerging bioremediation tool with unique broad spectrum methanotrophs. In: Agro-environmental sustainability (two volumes from Springer-2017). https://doi.org/10.1007/978-3-319-49727-3_1

Shraddha, Shekher R, Sehgal S, Kamthania M, Kumar A (2011) Laccase: microbial sources, production, purification, and potential biotechnological applications. Enzym Res 2011(1): 217861

Strong PJ, Claus H (2011) Laccase: a review of its past and its future in bioremediation. Crit Rev Environ Sci Technol 41:373e434. https://doi.org/10.1080/10643380902945706

Su FH, Tabanag IDF, Wu CY, Tsai SL (2017) Decorating outer membrane vesicles ~ with organophosphorus hydrolase and cellulose binding domain for organophosphate pesticide degradation. Chem Eng J 308:1e7

Thatoi H, Das S, Mishra J, Rath BP, Das N (2014) Bacterial chromate reductase, a potential enzyme for bioremediation of hexavalent chromium: a review. J Environ Manag 15(146):383–399. https://doi.org/10.1016/j.jenvman.2014.07.014

Weinberger N, Jörissen J, Schippl J (2012) Foresight on environmental technologies: options for the prioritisation of future research funding–lessons learned from the project "Roadmap Environmental Technologies 2020+". J Clean Prod 27:32–41

Wu Z, Kan SBJ, Lewis RD, Wittmann BJ, Arnold FH (2019) Machine learning-assisted directed protein evolution with combinatorial libraries. Proc Natl Acad Sci USA 116:8852–8858

Yang C, Xia Y, Qu H, Li AD, Liu R, Wang Y, Zhang T (2016) Discovery of new cellulases from the metagenome by a metagenomics-guided strategy. Biotechnol Biofuels 9(1):138

Yim SK, Jung HC, Pan JG, Kang HK, Ahn T, Yun CH (2006) Functional expression of mammalian NADPHecytochrome P450 oxidoreductase on the cell surface of Escherichia coli. Protein Expr Purif 49:292e298

Yin K, Lv M, Wang Q, Wu Y, Liao C, Zhan GW, Chen L (2016) Simultaneous bioremediation and detection of mercury ion through the surface display of carboxylesterase E2 from Pseudomonas aeruginosa PA1. Water Res 103:383e390. https://doi.org/10.1016/j.watres.2016.07.053

Zhanga H, Zhanga S, Hea F, Qina X, Zhanga X, Yang Y (2016) Characterization of a manganese peroxidase from white-rot fungus Trametes sp. 48424 with strong ability of degrading different types of dyes and polycyclic aromatic hydrocarbons. J Hazard Mater 320:265e277. https://doi.org/10.1016/j.jhazmat.2016.07.065

Chapter 4
Enzymes in Valorization of Waste: Future Advancement Through the Biotechnological Revolution

Abstract The depletion of fossil fuel supplies, global warming, unabated population increase, and expensive and troublesome waste recycling necessitate using renewable energy and consumer goods. It is possible to create biomass-based industrial methods as an alternative to the 100% oil economy. Globally, enormous volumes of lignocellulosic wastes are generated each year. These consist of food farming residues, tree pruning residues, "greengrocer's wastes," agricultural residues, and urban solid waste's organic and paper components. The prevalent methods used to dispose of this trash negatively affect the environment and the economy. As an alternative, procedures for upgrading wastes into high-value goods with economic and ecological benefits should be created. This is known as the upgrading concept, which adds value to waste by producing a product with desirable, repeatable features. Upgrading solid wastes can yield a wide range of high-value products, including organic acids, biofuels, enzymes, biopolymers, bioelectricity, and compounds for the food and pharmaceutical industries. This manuscript outlines the latest developments in biotechnological procedures for their manufacturing.

Keywords Enzymes · Biofuels · Organic acids · Food waste · Valorisation · Sustainability

4.1 Introduction

The depletion of fossil fuel supplies, global warming, unabated population increase, and expensive and troublesome waste recycling necessitate using renewable energy and consumer goods. Utilizing the entire plant, production methods based on biomass can be developed as an alternative to the 100% oil economy, resulting in the biorefinery concept. It is possible to extract and functionalize every component of the leftover biomass to create synthons, intermediate agro-industrial products, and food and non-food fractions. The three main industrial realms of materials, energy, and molecules can all be affected. Reducing externalities related to waste disposal would be one benefit of using residual biomass as a raw source.

A generalization of the biorefinery concept using the entire plant is possible. The reduction of wastes or byproducts may be related to transforming a single plant,

© The Author(s), under exclusive license to Springer Nature Switzerland AG 2024   57
J. A. Parray et al., *Enzymes in Environmental Management*, SpringerBriefs in Environmental Science, https://doi.org/10.1007/978-3-031-74874-5_4

using multiple plants based on the complementary qualities of various processes, distinct factories that occasionally belong to different firms or even the zero-waste biorefinery concept. The first one's wastes and byproducts may serve as a source of energy for the third and a starting material for the second. A generalization of the idea of a biorefinery through an actual "Industrial Metabolism" is the optimization of raw materials, byproducts, and energy between various industrial production units on a specific site (Octave and Thomas 2009). Oil-based products replaced by biobased ones will establish a new economy and industrial processes that adhere to sustainable development. Developing novel procedures based on the 12 principles of green chemistry—clean processes, atom economy, renewable feedstocks, etc.—will be connected to industrial biorefineries. Biotechnology—mainly white biotechnology—will play a significant role in these novel processes involving fermentation and biotransformations (enzymology and microbes). Globally, enormous volumes of lignocellulosic wastes are generated each year. These consist of food farming residues, tree pruning residues, "greengrocer's wastes," agricultural residues, and urban solid waste's organic and paper components. Currently, the most popular methods for getting rid of these wastes are feeding them to animals, which has the drawback that not all wastes are suitable for all animals, incineration, which is inconvenient due to its high water content and low calorific value, and landfilling, which has the significant drawback of being very expensive to transport. Alternatively, procedures for upgrading wastes—converting wastes into products with desired qualities and benefits for the economy and environment—should be created (Laufenberg et al. 2003). An extensive range of high-value products can be obtained by upgrading solid wastes through biotechnological processes, including enzymes, biofuels, organic acids, biopolymers, bioelectricity, and compounds for the food and pharmaceutical industries. This document provides an outline of the latest developments in their creation.

4.2 Food Waste Valorisation Techniques

Food waste valorization methods aim to transform food waste into products with added value that can be included in different phases of the food supply chain. The following are some environmentally friendly methods for valorizing food waste: anaerobic digestion, composting, bioconversion, and upcycling. Food waste is abundant in proteins, carbohydrates, lignin, lipids, cellulose, and other elements and can be found in solid, semi-solid, and liquid forms. These compounds are typically found in diluted form in liquid food waste, such as whey from the dairy industry, potato processing wastewater, and apple residue sludge. Food waste is categorized as an affordable, high-potential second-generation raw resource because of its rich composition. The composition of the generated food waste is the cornerstone of every process for valorizing food waste (Carmona-Cabello et al. 2020). The composition of food waste varies and is contingent upon a given population's geography, season, and dietary habits. According to Kazemi et al. (2022), heterogeneous food waste typically contains less than 2% of cellulose, over 50% of total sugars, over 40% of starch, and 45% of proteins. In addition to many other uses, the materials found in food waste can be used to create biofuels, bulk chemicals, enzymes, food supplements, nutraceuticals, bio-adsorbents (like activated charcoal), antimicrobial products (like antibiotics), bio-based fertilizers, and bioplastics (O'Connor et al. 2021). This section addresses the recovery of resources from food waste by applying several sustainable methods or strategies that seek to transform food waste into products with added value that can be incorporated into different phases of the food supply chain. Composting, anaerobic digestion, bioconversion, and upcycling to value-added derivatives and new food products/ingredients like fertilizer, animal feed, and food additives are examples of food waste valorization techniques that are considered sustainable in the waste management literature (Nayak and Bhushan 2019).

4.2.1 Composting

In a controlled process known as food waste composting, food wastes are broken down into their most basic components by a sophisticated microbial population, either with or without oxygen (Amuah et al. 2022). The four stages of composting include maturation, cooling, thermophilic, and mesophilic. The carbon-nitrogen (C/N) ratio, pH, moisture, feedstock nature (kind and composition of food waste), oxygen availability (aeration), and composting method are some of the elements that affect how long composting takes (Meena et al. 2021). Compared to anaerobic composting, which can take up to two or three years, aerobic composting is the most efficient form of decomposition and yields mature compost in a shorter amount of time (4 weeks). In addition to using effective microbial species, chemical nitrogen activators, worms, natural minerals, plant-based bulking agents (such as agricultural waste), and a variety of additives and amendments, such as biochar, composting can be accelerated by regularly turning the compost pile and shredding the feedstock materials (i.e., food waste) (Waqas et al. 2018). Compost is a stabilized material rich in organic materials resulting from composting. Compost is used in agriculture because it increases soil fertility and fosters crop development and growth. It has been used as an organic fertilizer as a substitute for chemical fertilizers, whose excessive and unchecked use has, over time, had a detrimental impact on agricultural production and soil quality indicators. It functions as a natural soil conditioner as well. Humic compounds and nutrients abound in compost. Humic substances are heterogeneous organic molecules that include human, fulvic acids, and humic acids (HAs). Humic chemicals improve soil physiochemical properties, structure, and texture and raise soil fertility (Palanivell et al. 2013).

Additionally, adding compost from food waste that has undergone biotransformation to soils improves soil aggregation, water-retentiveness, and porosity (Das et al. 2024). Compost's humic materials positively affect plants' physiological, metabolic, and developmental processes. Both the plasma membrane H + -ATPase enzyme, which is involved in nutrient uptake and transport, and genes linked to nitrate uptake in plants are activated by humic compounds. Studies show that humic substances promote stomatal opening, which is required for respiration and photosynthesis. In addition to improving the composition of soil organic matter, they play a significant role in the restoration of land resources through phytoremediation and vegetation revival in deficient arable soils (Chen et al. 2022).

Composting has several advantages but drawbacks, such as lengthy mineralization times, protracted processing times, the possibility of thermotolerant pathogens, and low nutritional value. There are also drawbacks to this biological waste treatment method, including the production of chemicals and odorous substances (such as hydrogen sulfides generated in anaerobic circumstances) as well as greenhouse gases (CO_2, N_2O, and CH_4) (Amuah et al. 2022). The availability of nutrients to plants, primarily N and P, is unknown because they exist in both inorganic and organic forms and are not readily ingested by crops. This is one of the main limitations of using compost as a biofertilizer. However, because compost releases nutrients over time, fewer applications of compost are required to achieve the same nutritional load as mineral fertilizers. Less than eight times as much of this compost (1% N) would be needed over four years to supply the same amount of nitrogen as mineral fertilizer. However, as compost is a variety of nutrients, it will be very beneficial to soil health to extract some (such as N, P, K, etc.) but not all. When applied to the soil for bioremediation, trace elements, such as those found in matured compost, may be added to the already present trace elements rather than removed, which could harm the soil's quality.

4.2.2 Anaerobic Digestion

The breakdown of organic molecules by microbes in an oxygen-free environment is known as anaerobic digestion. This process produces biogas and is helpful as an energy source. According to Uddin et al. (2021), hydrolysis, acidogenesis, acetogenesis, and methanogenesis are the processes involved in anaerobic digestion. Food waste can be anaerobically digested to provide renewable energy on-site and reduce methane emissions at landfills, among many other benefits. As a byproduct of anaerobic digestion, the food waste substrate is removed to yield a digestate rich in phosphate and nitrogen (Song et al. 2021). Because organic fertilizers, such as food waste anaerobic digestate, are rich in organic matter, a long-term release of nutrients can be achieved by decomposing delicate organic matter maintained within the cultivated soil after fertilizer application. The amount of nutrients plants can eventually take is limited by dissolved synthetic fertilizers, making nutrients accessible in their formulations susceptible to nutrient leaching and fixing (Cheong et al. 2020). Anaerobic digestion and other organic fertilizers improve the fertility and characteristics of the soil. In contrast, chemical fertilizers only supply nutrients and have no beneficial effect on the physical state of the soil.

Furthermore, according to some studies, correctly applying digestate can maintain crop output while halting the spread of diseases carried by the soil and perhaps reducing nutrient runoff. Applying this waste treatment method as a soil biofertilizer performs better than using regular compost. For the treatment and disposal of high-moisture biomass, such as food waste, anaerobic digestion is preferred over composting and landfilling (Dutta et al. 2021) because of its reduced emissions of greenhouse gases and smells. Furthermore, using anaerobic digestion on a wide scale may be financially beneficial in producing high-value products like bio-CNG (biogas) (Lin et al. 2018). Even though anaerobic digestion has the benefits mentioned above, food waste is usually a source of pathogenic microbes. Some pathogens, like spore-formers and Listeria, can survive the anaerobic digestion process and remain in the digestate, posing a risk to the environment and potentially causing severe illnesses in humans and animals. Inaccurate biofertilizer application and unbalanced nutrient release can be caused by miscalculating fertilizer dose and availability, which may jeopardize crop development and soil quality (Guruchandran et al. 2022). The digestate's high water content presents problems and raises storage and transportation expenses. Thus, there is a need to improve anaerobic digestate's applicability as a biobased fertilizer. Furthermore, a significant obstacle to adopting this technology is the high capital cost associated with anaerobic digestion (Lin et al. 2018).

4.2.3   Bioconversion

The process of transforming organic resources, such as biomass from animal or vegetative waste, into marketable goods or renewable energy sources by biological processes or agents, such as microbial species, enzymes, etc., is known as biotransformation, sometimes known as bioconversion (Maurya et al. 2021). To improve food security and support sustainable development, food wastes can be bioconverted into value-added products such as feed and food additives, bio-based fuels, animal feeds, enzymes, food-grade pigments, fertilizers, single-cell protein (SCP), and other valuable biochemicals or bioproducts (Tropea 2022). According to Hashempour-Baltork et al. (2022), bioconversion offers several benefits, such as high biomaterial and energy recovery value, a decrease in landfill space, cheaper technology costs, and an increase in farmer revenue from the sale of agricultural and horticulture waste. Several scientists investigated the possibility of using food waste as an alternate source of raw materials to produce goods that would be beneficial. For instance, generating single-cell protein through fermentative bioconversion has shown enormous benefits. It is meant to solve global protein shortfalls as it is not dependent on soil health, climate, or cultivable land availability (Tropea et al. 2022). During fermentation, food waste is broken down into simpler molecules by microbial cells such as bacteria, fungi, yeast, and algae. This produces high-quality protein sources for fish, animals, and humans to consume. This process is used to make single-cell proteins. According to LaTurner et al. (2020), SCP can be a cost-effective alternative to traditional protein sources like soy meal and protein-based additives used in animal feed formulations like fishmeal and meat meal. Studies have shown that waste from oranges, lemons, corn stover, pineapple, mango, banana, apple, cashew apple, jackfruit, cacao, pomegranate, cheese, coffee (Abdullahi et al. 2021), fish, and pineapple can produce single-cell proteins (Mahan et al. 2018), as well as bananas (Nisansala et al. 2021). However, there are still regulatory barriers and questions over the goods' safety of potentially dangerous or anti-nutritional ingredients. Furthermore, to avoid unfavorable consequences and burden-shifting, it is imperative to evaluate the environmental sustainability of these waste-to-nutrition processes as early as possible when aiming to replace conventional food constituents like proteins (Siegrist et al. 2023).

Starmerella bombicola ATCC 22214 was used to bioconvert restaurant leftovers and bakery trash into sophorolipids through a fed-batch fermentation procedure and in-situ separation. Even though this approach demonstrated a high level of efficiency, more industrial-scale separation process optimization might be implemented. Biodegradable sophorolipids have a lot of potential for use in agriculture and food industries as a substitute for chemical surfactants (Madankar and Borde 2023). Potential uses include developing bio-preservatives and active packaging to improve food safety because of their potent anti-bacterial qualities as well as raising crop yield by improving the solubility and mobility of nutrients (Celligoi et al. 2020). Using Acetobacter tropicalis KFCC 11476P, fed-batch fermentation was also used to create vinegar from leftover onion juice (Saha et al. 2023). For the bioconversion of agro-industrial food waste, fermentation techniques—in particular, solid-state fermentation (SSF)—have been extensively studied and found to be an efficient process that yields affordable enzymes, organic acids, vitamins, fermented feeds, and fragrance compounds (Hadj Saadoun et al. 2021). According to Sharma et al. (2022), SSF has the following benefits: cheaper production costs, more productivity, less waste created, easier equipment design, and organic culture media made from solid food and agricultural waste. SSF has several advantages over other fermentation techniques, such as increased product stability and efficiency, better product activity and quality, enhanced end-product recovery, and the lack of extraction solvents. Furthermore, according to Yao et al. (2021), SSF is a productive technique for breaking down insoluble macromolecules and anti-nutritional factors, which enhances nutrient absorption and utilization. Saccharomyces cerevisiae, Candida utilis, and Yarrowia lipolytica were used in Li et al.'s (Saha et al. 2023) study to bioconvert food waste from a canteen into fermented feed for crayfish aquaculture. According to the study, crayfish growth can be supported by supplementing 30% of the fermented feed from food waste with the basal diet (Li et al. 2023). Using a medium including sugarcane molasses, SSF was previously studied to synthesize the packaging material poly-3-hydroxybutyrate (PHB) (McAdam et al. 2020). Furthermore, five distinct microbial consortiums were investigated for their ability to create PHB from food and kitchen waste. The greatest PHB concentration (0.91 g/L) was produced by Consortium A, which was made up of three different Bacillus species (B. sonorensis MPTD1, B. cereus MPTDC, and B. pumilus MPTDFH). While SSF is an appealing technology, a few issues must be resolved. For example, a slower rate of microbial growth results from the low moisture content of the waste substrates utilized in SSF, which also limits enzymatic activity and restricts the movement of nutrients and metabolites. Furthermore, only microorganisms capable of surviving in low moisture environments can be used with SSF (Yang et al. 2021). Insect-based bioconversion is a relatively new waste valorization technique investigated recently for producing valuable products when insects break down food waste. Value can be created at various stages of this process. For instance, selling insect-based biomass as a source of protein for food and feed, commercializing fractionated secondary products (i.e., biomolecules), and selling the remaining bio-converted byproducts as soil amendments for horticulture and agriculture are some examples of how eliminating food waste can reduce disposal costs (Fowles and Nansen 2020). On the other hand, one of the difficulties faced by insect-based food waste valorization processes is the consumer acceptability of the created items. Food waste can be used to feed insects, which has various advantages for the environment and the economy. First, insects, such as flies, crickets, and grasshoppers, subsist on food scraps that people and animals cannot eat. Second, by feeding insects with food that would otherwise go to waste, individuals can incorporate insects into their meals and contribute to developing a bioeconomy. Additionally, as Ojha et al. (2020) noted, this would support preserving a balance between the global supply and demand for animal protein. The black soldier flies, Hermetia illucens, are great insects for bioconversion. This insect reproduces quickly, and its larvae have an insatiable appetite, a high food conversion rate, and the ability to consume a wide range of organic waste, including food waste (Gao et al. 2019). Because black soldier fly larvae (BSFL) have high nutritional value, so they have drawn much attention. Lipids, proteins, minerals, vitamins, and vital amino acids are present in high concentrations in them (Lu et al. 2022). Because they are high in protein and can help meet the growing need for protein, they are suitable candidates for inclusion in humans' diets as edible insects. Nonetheless, Bessa et al. (2021) contended that several current issues impede the acceptance of BSFL as a wholesome food source for human consumption. Although consumers are prepared to eat processed food from BSFL, little is known about its safety because it has primarily been utilized as animal feed.

Despite these benefits, raising insects can result in large amounts of low-value organic waste, such as frass, exuviae, and uneaten feed, frequently spread over farmlands for disposal (Ojha et al. 2020). Researchers have voiced their worries and suggested more research before introducing large-scale commercial insect farming and valorization to prevent potential risks. This is because certain insects include harmful compounds, including cyanide or thiaminases, that make them less beneficial to humans. Thus, research on insects' nutritional value and safety for food and feed must complement ecological studies on selecting safe species for domestication and eating. Insects that escape from insect farms have also been cited as a potential hazard to environmental equilibrium.

4.3   Added Value Products from Wastes by Microbial Solid-State Fermentation

Solid-state fermentation (SSF) is characterized by low energy consumption, high volumetric productivity, higher titer of added-value products, minimal waste formation, and low catabolic repression. It is conducted in the absence or near absence of free water (Hölker et al. 2004). The following reports on effectively utilizing several waste types as substrates for microbial solid-state fermentation to produce a range of products with significant cost advantages. To enable microbial colonization, solid wastes must undergo a simple pretreatment process that includes material categorization in various granulometries and grinding to improve material homogenization and ensure that this parameter has less of an impact on the subsequent stage. Thus, SSF is a compelling substitute for the Submerged Fermentation (SmF) method. The use of different organic wastes with their potential to produce bio-products is shown in Table 4.1.

Table 4.1 List of organic wastes and their potential use in solid-state fermentation (SSF) to obtain value-added bio-products

Category of organic waste	Type of products/ processes	Waste materials/ residues	potential use	References
Municipal/ domestic food waste	Kitchen waste	Preparation waste, left-over food, sludge, cocoyam peels	Biopesticides, animal feedstuff, organic acids, antibiotics	Zhang et al. (2015)
	Commercial/ hotel	Used coffee grounds, used tea bags, wasted bread, leftovers, expired foods, and sludge.	Enzymes, animal feed-stuff, biopesticides, bioethanol, bioplastics	Rocha et al. (2014)
Industrial organic waste	Sugar industry	Molasses	Enzymes, fructooligosaccharides	Ghazi et al. (2006)
	Poultry processing	Skin, blood, fats, hairs, feathers, bones, liver, intestines, wings, trimmed organs	Enzymes, animal feed-stuff, biofertilizers	Jayathilakan et al. (2012)
	Marine products processing	Shells, roes, pincers, trimmed parts	Enzymes, bioactive compounds	Kandra et al. (2012)
	Fruits and vegetables processing	Skin, peels, pomace, fiber, kernel, stones, seeds	Pectinolytic enzymes, biofuels, animal feed, organic acids	Panda et al. (2016)
Agricultural organic waste	Cattle, broiler	Fleshing wastes, dung, litter	Animal glue, animal feed supplement, methane production, biochar, activated carbon	Adams et al. (2002)
	Oils and oil seeds	Shells, husks, fibers, sludge, press cake	Bioethanol, enzymes, biofuels, biofertilizers	Jørgensen et al. (2010)

4.3.1   The Challenges of Implementing Food Waste Valorisation

Putting food waste valorization ideas into practice presents many intricate and varied problems. These tasks can be broken down into multiple categories.

4.3.2   Financial Constraints and Resistance to Change

Adopting food waste valorization technologies is needed because of a lack of funding and aversion to change (Martin-Rios et al. 2021). Many organizations may be discouraged from implementing these technologies since they frequently require significant financial commitment. Furthermore, adopting novel valorization processes may need to be improved by opposition to departing from current practices and methodologies. Other issues that require attention are high processing costs and overall process viability. Food waste valorization may not be as cost-effective as it may be due to the expenses of traditional extraction methods, waste transportation and storage, and other related factors. These issues make it challenging for valorized products to compete with conventional alternatives.

4.3.3   Technical and Regulatory Obstacles

Additional barriers to successful food waste valorization include the technical challenges of transforming extracted materials into value-added products, particularly on a big scale (Socas-Rodríguez et al. 2021). Creating compelling and scalable procedures is necessary to increase the conversion of extracted chemicals into commercially viable products. The application of valorization solutions is further complicated by the absence of a robust legislative framework for handling trash.

4.3.4   Heterogenous Composition of Food Waste

According to Makanjuola et al. (2020), there are notable seasonal, geographical, and nutritional differences in food waste composition. It isn't easy to decide which valorization technology is most suited for scaling up because of these variances. To efficiently use food waste in valorization processes, clever strategies to address these composition inconsistencies must be developed (Hu et al. 2021).

4.3.5 Unavailability of Food Waste Generation Data and Quantification Difficulties

Data on food waste production is scarce, especially in developing nations. Our comprehension of the regional differences in food waste, the resulting nutrient loss, and the ecological ramifications is hampered by this knowledge gap. Food waste creation must be identified and quantified to choose the best valorization strategies and pinpoint the primary resources needed to process value-added products. Targeted reduction efforts can be facilitated by identifying hotspots of food losses and waste formation. This can help prioritize food supply chain stages and waste categories with the most significant valorization potential (Guo et al. 2014). The definition of food waste and losses might change depending on the conceptual frameworks and policy targets. These differences may impact food waste detection and quantification techniques (Cattaneo et al. 2021). To ensure the sustainability and effectiveness of waste valorization technologies, consideration must be given to regional needs and perspectives during the development process (Hu et al. 2021).

4.3.6 Quality and Safety Challenges

It is essential to guarantee the safety and quality of valorized waste and byproducts. Food waste may contain pathogens, organic residues, chemicals, and water that could harm one's health. To provide guidelines and standards for the secure and superior valorization of waste, laws, rules, and research are required. Furthermore, for widespread adoption, it is imperative to address consumer concerns about the acceptability and safety of valorized products through focused education and awareness campaigns (Fanelli and Di Nocera 2017). In conclusion, putting food waste valorization techniques into practice requires overcoming several obstacles, such as lack of funding, opposition to change, high processing costs, technological issues, regulatory frameworks, compositional differences, data shortages, and safety concerns. A multidisciplinary approach, industry and public-private partnership engagement, and research and development funding are essential to overcome these obstacles and develop long-term strategies for valorizing food waste (Al-Obadi et al. 2022).

4.4 Conclusions

Food loss and waste are worldwide issues that require thorough and focused solutions. Even if a growing number of stakeholders are becoming more engaged in this subject, it is alarming that the production of food waste is increasing, which is contributing to a rise in hunger and malnutrition on a global scale. To solve these problems, more food must be produced in a way that is safe for the environment and efficient in handling food waste. It is imperative to implement mass-scale food waste valorization strategies to accomplish these aims. By highlighting the circular economy as a central innovation concept, waste valorization techniques might provide more benefits, generating revenue from food waste and losses. This strategy takes advantage of the nutritious qualities and composition of food waste and inedible items to facilitate their reintegration into the food and agriculture industries. This helps to promote food security and fight hunger. The study emphasizes the encouraging potential of programs to value food waste and create goods that may be used again in agricultural and food production processes.

It is imperative to recognize that food waste's content and environmental effects exhibit national variations, hence requiring customized waste reduction approaches for every nation. The market for waste-to-value products is anticipated to grow as consumer awareness rises, propelling technological developments in food waste valorization and improving customer perception. More funding and assistance are required to fully reap the rewards of food waste valorization and attain zero waste in food production and consumption. To maximize food waste's valorization, research should concentrate on tackling safety issues and the diverse composition of food waste. The ability to turn food waste into high-quality food-grade items will be limited without sufficient study, which will impede the movement toward ending world hunger. Lastly, it's critical to rank particular concerns about food loss and waste in the conclusions. We can drastically cut down on food waste, guarantee food security, and end hunger by implementing focused, research-driven ideas, such as large-scale valorization projects and customized waste reduction techniques. Technological developments, public awareness campaigns, and more investment will be the main forces behind the food waste valorization industry's growth.

References

Abdullahi N, Dandago MA, Yunusa AK (2021) Review on production of single-cell protein from food wastes. Turk J Agric Food Sci Technol 9:968–974

Adams TT, Eiteman MA, Hanel BM (2002) Solid state fermentation of broiler litter for production of biocontrol agents. Bioresour Technol 82:33–41

Al-Obadi M, Ayad H, Pokharel S, Ayari MA (2022) Perspectives on food waste management: prevention and social innovations. Sustain Prod Consum 31:190–208

Amuah EEY, Fei-Baffoe B, Sackey LNA, Douti NB, Kazapoe RW (2022) A review of the principles of composting: understanding the processes, methods, merits, and demerits. Org Agric 12:547–562. [CrossRef]

Bessa LW, Pieterse E, Marais J, Dhanani K, Hoffman LC (2021) Food safety of consuming black soldier fly (Hermetia illucens) larvae: microbial, heavy metal and cross-reactive allergen risks. Food Secur 10:1934

Carmona-Cabello M, García IL, Sáez-Bastante J, Pinzi S, Koutinas AA, Dorado MP (2020) Food waste from restaurant sector—characterization for biorefinery approach. Bioresour Technol 301:122779. [CrossRef] [PubMed]

Cattaneo A, Sánchez MV, Torero M, Vos R (2021) Reducing food loss and waste: five challenges for policy and research. Food Policy 98:101974

Celligoi MAPC, Silveira VAI, Hipólito A, Caretta T, Baldo C (2020) Sophorolipids: a review on production and perspectives of application in agriculture. Span J Agric Res 18:e03R01

Chen Q, Qu Z, Ma G, Wang W, Dai J, Zhang M, Wei Z, Liu Z (2022) Humic acid modulates growth, photosynthesis, hormone and osmolytes system of maize under drought conditions. Agric Water Manag 263:107447

Cheong JC, Lee JTE, Lim JW, Song S, Tan JKN, Chiam ZY, Yap KY, Lim EY, Zhang J, Tan HTW et al (2020) Closing the food waste loop: food waste anaerobic digestate as fertilizer for the cultivation of the leafy vegetable, Xiao Bai Cai (Brassica rapa). Sci Total Environ 715:136789

Das A, Verma M, Mishra V (2024) Food waste to resource recovery: a way of green advocacy. Environ Sci Pollut Res 31:17874–17886

Dutta S, He M, Xiong X, Tsang DCW (2021) Sustainable management and recycling of food waste anaerobic digestate: a review. Bioresour Technol 341:125915

Fanelli R, Di Nocera A (2017) How to implement new educational campaigns against food waste: an analysis of best practices in European countries. Econ Agro Aliment 19:223–244

Fowles TM, Nansen C (2020) Insect-based bioconversion: Value from food waste. In: Närvänen E, Mesiranta N, Mattila M, Heikkinen A (eds) Food Waste Management: Solving the Wicked Problem. Springer, Cham, pp 321–346. ISBN 978-3-030-20561-4

Gao Z, Wang W, Lu X, Zhu F, Liu W, Wang X, Lei C (2019) Bioconversion performance and life table of black soldier fly (Hermetia illucens) on fermented maize straw. J Clean Prod 230: 974–980

Ghazi I, Fernandez-Arrojo L, Gomez De Segura A, Alcalde M, Plou FJ, Ballesteros A (2006) Beet sugar syrup and molasses as low-cost feedstock for the enzymatic production of fructo-oligosaccharides. J Agric Food Chem 54:2964–2968

Guo W, Wu Q, Yang S, Luo H, Peng S, Ren N (2014) Optimization of ultrasonic pretreatment and substrate/inoculum ratio to enhance hydrolysis and volatile fatty acid production from food waste. RSC Adv 4:53321–53326

Guruchandran S, Muninathan C, Ganesan ND (2022) Novel strategy for effective utilization of anaerobic digestate as a nutrient medium for crop production in a recirculating deep water culture hydroponics system. Biomass Conv Bioref 14:1–13

Hadj Saadoun J, Bertani G, Levante A, Vezzosi F, Ricci A, Bernini V, Lazzi C (2021) Fermentation of agri-food waste: a promising route for the production of aroma compounds. Food Secur 10:707

Hashempour-Baltork F, Farshi P, Khosravi-Darani K (2022) Vegetable and fruit wastes as substrate for production of single-cell protein and Aquafeed meal. In: Ray RC (ed) Fruits and vegetable wastes: Valorization to bioproducts and platform chemicals. Springer, Singapore, pp 169–187. ISBN 9789811695278

Hölker U, Höfer M, Lenz J (2004) Biotechnological advantages of laboratory-scale solid-state fermentation with fungi. Appl Microbiol Biotechnol 64:175–186

Hu X, Subramanian K, Wang H, Roelants SLKW, Soetaert W, Kaur G, Lin CSK, Chopra SS (2021) Bioconversion of food waste to produce industrial-scale sophorolipid syrup and crystals: dynamic life cycle assessment (DLCA) of emerging biotechnologies. Bioresour Technol 337: 125474

Jayathilakan K, Sultana K, Radhakrishna K, Bawa AS (2012) Utilization of byproducts and waste materials from meat, poultry and fish processing industries: a review. J Food Sci Technol 49: 278–293

Jørgensen H, Sanadi AR, Felby C, Lange NEK, Fischer M, Ernst S (2010) Production of ethanol and feed by high dry matter hydrolysis and fermentation of palm kernel press cake. Appl Biochem Biotechnol 161:318–332

Kandra P, Challa MM, Jyothi KP, H. (2012) Efficient use of shrimp waste: present and future trends. Appl Microbiol Biotechnol 93:17–29

Kazemi SPH, Dehhaghi M, Guillemin GJ, Gupta VK, Lam SS, Aghbashlo M, Tabatabaei M (2022) Bioethanol production from food wastes rich in carbohydrates. Curr Opin Food Sci 43:71–81. [CrossRef]

LaTurner ZW, Bennett GN, San K-Y, Stadler LB (2020) Single cell protein production from food waste using purple non-sulfur bacteria shows economically viable protein products have higher environmental impacts. J Clean Prod 276:123114

Laufenberg G, Kunz B, Nystroem M (2003) Transformation of vegetable waste into value added products: (A) the upgrading concept; (B) practical implementations. Bioresour Technol 87: 167–198

Li Q, Yi P, Zhang J, Shan Y, Lin Y, Wu M, Wang K, Tian G, Li J, Zhu T (2023) Bioconversion of food waste to crayfish feed using solid-state fermentation with yeast. Environ Sci Pollut Res 30: 15325–15334

Lin L, Xu F, Ge X, Li Y (2018) Improving the sustainability of organic waste management practices in the food-energy-water nexus: a comparative review of anaerobic digestion and composting. Renew Sust Energ Rev 89:151–167

Lu S, Taethaisong N, Meethip W, Surakhunthod J, Sinpru B, Sroichak T, Archa P, Thongpea S, Paengkoum S, Purba RAP et al (2022) Nutritional composition of black soldier fly larvae (Hermetia illucens L.) and its potential uses as alternative protein sources in animal diets: a review. Insects 13:831

Madankar C, Borde P (2023) Review on Sophorolipids—a promising microbial bio-surfactant. Tenside Surfactant Deterg 60:95–105

Mahan KM, Le RK, Wells T, Anderson S, Yuan JS, Stoklosa RJ, Bhalla A, Hodge DB, Ragauskas AJ (2018) Production of single cell protein from agro-waste using rhodococcus opacus. J Ind Microbiol Biotechnol 45:795–801

Makanjuola O, Arowosola T, Du C (2020) The utilization of food waste: challenges and opportunities. J Food Chem Nanotechnol 6:182–188

Martin-Rios C, Hofmann A, Mackenzie N (2021) Sustainability-oriented innovations in food waste management technology. Sustain For 13:210

Maurya DK, Kumar A, Chaurasiya U, Hussain T, Singh SK (2021) 11—Modern Era of microbial biotechnology: Opportunities and future prospects. In: Solanki MK, Kashyap PL, Ansari RA, Kumari B (eds) Microbiomes and Plant Health. Academic, Cambridge, MA, pp 317–343. ISBN 978-0-12-819715-8

McAdam B, Brennan Fournet M, McDonald P, Mojicevic M (2020) Production of polyhydroxybutyrate (PHB) and factors impacting its chemical and mechanical characteristics. Polymers 12:2908

Meena A, Karwal M, Dutta D, Mishra RP (2021) Composting: phases and factors responsible for efficient and improved composting. Agric Food E Newsl 1:85–90

Nayak A, Bhushan B (2019) An overview of the recent trends on the waste valorization techniques for food wastes. J Environ Manag 233:352–370

Nisansala A, Nikagolla D, Kapilan R (2021) Fruit waste substrates to produce single-cell proteins as alternative human food supplements and animal feeds using Baker's yeast (*Saccharomyces cerevisiae*). J Food Qual 2021:9932762

O'Connor J, Hoang SA, Bradney L, Dutta S, Xiong X, Tsang DCW, Ramadass K, Vinu A, Kirkham MB, Bolan NS (2021) A review on the valorisation of food waste as a nutrient source and soil amendment. Environ Pollut 272:115985. [CrossRef]

Octave S, Thomas D (2009) Biorefinery: toward an industrial metabolism. Biochimie 91:659–664

Ojha S, Bußler S, Schlüter OK (2020) Food waste valorisation and circular economy concepts in insect production and processing. Waste Manag 118:600–609

Palanivell P, Susilawati K, Ahmed OH, Majid NM (2013) Compost and crude humic substances produced from selected wastes and their effects on Zea mays L. nutrient uptake and growth. Sci World J 2013:e276235

Panda SK, Mishra SS, Kayitesi E, Ray RC (2016) Microbial-processing of fruit and vegetable wastes for production of vital enzymes and organic acids: biotechnology and scopes. Environ Res 146:161–172

Rocha MVP, de Matos LJBL, De Lima LP, Figueiredo PMDS, Lucena IL, Fernandes FAN, Gonçalves LRB (2014) Ultrasound-assisted production of biodiesel and ethanol from spent coffee grounds. Bioresour Technol 167:343–348

Saha D, Das PK, Saha D, Das PK (2023) *Bioconversion of agricultural and food wastes to vinegar*. IntechOpen, Rijeka. ISBN 978-1-80356-849-2

Socas-Rodríguez B, Alonso-García DM, Mendiola JA, Cifuentes A, Ibáñez E (2021). Betaine-based natural deep eutectic solvents as promising green extraction agents for pesticides determination in valorized food by-products

Sharma V, Tsai M-L, Nargotra P, Chen C-W, Kuo C-H, Sun P-P, Dong C-D (2022) Agro-industrial food waste as a low-cost substrate for sustainable production of industrial enzymes: a critical review. Catalysts 12:1373

Song S, Lim JW, Lee JTE, Cheong JC, Hoy SH, Hu Q, Tan JKN, Chiam Z, Arora S, Lum TQH et al (2021) Food-waste anaerobic digestate as a fertilizer: the agronomic properties of untreated digestate and biochar-filtered digestate residue. Waste Manag 136:143–152

Tropea A (2022) Food waste valorization. *Fermentation* 8:168

Tropea A, Ferracane A, Albergamo A, Potortì AG, Lo Turco V, Di Bella G (2022) Single cell protein production through multi food-waste substrate fermentation. *Fermentation* 8:91

Uddin MN, Siddiki SYA, Mofijur M, Djavanroodi F, Hazrat MA, Show PL, Ahmed SF, Chu Y-M (2021) Prospects of bioenergy production from organic waste using anaerobic digestion technology: a mini review. Front Energy Res 9:627093

Waqas M, Nizami AS, Aburiazaiza AS, Barakat MA, Ismail IMI, Rashid MI (2018) Optimization of food waste compost with the use of biochar. J Environ Manag 216:70–81. [CrossRef]

Yang L, Zeng X, Qiao S (2021) Advances in research on solid-state fermented feed and its utilization: the pioneer of private customization for intestinal microorganisms. Anim Nutr 7: 905–916

Yao Y, Li H, Li J, Zhu B, Gao T (2021) Anaerobic solid-state fermentation of soybean meal with *Bacillus* sp. to improve nutritional quality. Front Nutr 8:706977

Zhang W, Zou H, Jiang L, Yao J, Liang J, Wang Q (2015) Semi-solid state fermentation of food waste for production of Bacillus thuringiensis biopesticide. Biotechnol Bioprocess Eng 20: 1123–1132

Chapter 5
Enzymes Technology in Biofuel Production

Abstract Biofuels are organic materials that can effectively replace fossil fuels to protect the environment. Various biofuels, including biodiesel, biohydrogen, bioethanol, and biogas, are produced today through specific production pathways. The production of these different biofuels depends on enzymes, either for the pretreatment of feedstock or as essential components of the synthesis process; the mode of action of these enzymes determines the feasibility and efficiency of the process. Commercializing biofuels from variable biomass has certain limitations related to the enzymes involved. Appropriate comprehension of the method of action offers answers for both breakthroughs and restrictions. Using genetic engineering techniques opens up new possibilities for advancement and makes the enzymatic process more affordable. The review addressed the state of enzymes in biofuel production processes today and relevant advancements in research and development.

Keywords Biofuel · Enzymes · Biodiesel · Biohydrogen · Bioethanol · Biogas

5.1 Introduction

Biofuels are one of the many resources available to meet the energy demand. Biomass is the term for the organic resources used to make biofuel. Animal, plant, algal, and industrial waste are the biomass sources used to generate biofuels. According to Singh et al. (2022), several methods for converting biomass include pyrolysis, anaerobic digestion, gasification, fermentation, and traditional burning. India ranked fourth among petroleum-consuming nations, behind the United States, China, and Japan, with an annual impact on economic standing of 6–8%. This places a heavy reliance on petroleum-based goods, such as kerosene, petrol, diesel, and natural gas, which have several dangers and difficulties, including pollution, global warming, health issues, waste production, etc. First-generation biofuels include ethanol, butanol, and propanol. Biofuels are classified into generations based on the fermentation of sugar and starch crops to produce alcohol. Cellulose, derived from waste biomass and non-food crops, is the raw material used to make second-generation biofuels, also known as cellulosic biofuels.

© The Author(s), under exclusive license to Springer Nature Switzerland AG 2024
J. A. Parray et al., *Enzymes in Environmental Management*, SpringerBriefs in Environmental Science, https://doi.org/10.1007/978-3-031-74874-5_5

Algal and microalgal species aid in synthesizing third-generation biofuels that benefit the world's population economically, socially, and environmentally while maintaining high energy efficiency for both industrial and individual consumption. Waste biomass includes livestock manure, municipal wastes, agricultural residues, and industrial organic wastes. Although waste biomass is less effective in poor nations for creating solid biofuel, they are still an excellent source of biomass generation from waste, such as orange peels used to produce biochar (Selvarajoo et al. 2022). A route to a low-carbon environment and a sustainable circular economy is made possible by using waste biomass to manufacture biofuels. It has been reported that Jatropha fruit waste (cake) is more suitable than rice waste and date fruit waste due to factors such as water content, percentage of volatile components and ashes, net calorific value (NCV), and essential chemical elements (C, H, N, and O) present in plant's waste biomass (Brunerová et al. 2017). Enzymes, the foundation of any biofuel company, assist the various biochemical processes that produce biofuels. According to Malode et al. (2022), these enzymes play a significant role in the production process, replacing traditional catalysts, cost-effectiveness, and an environmentally benign life cycle. Since most of these enzymes are driven by microorganisms, it is essential to have a thorough knowledge of the microbes and related enzymes that are unique to particular biofuels. Creating enzyme-based biofuel cells (EBCs), which use various nanomaterials to produce power from glucose as biofuel, is one of the most exciting recent advancements in biofuel technology (Kumar et al. 2018). It is necessary to have a deeper understanding of the enzymes responsible for biofuel generation since they are the primary determinant of any biofuel's cost-effectiveness and environmental quality. The current state of enzymes in the synthesis of biofuels and the advancement of technologies that have demonstrated some sustainable economic promise globally are the main topics of this chapter. Some of the advanced approaches for biofuel production are shown in Table 5.1.

5.2 Biofuel Option

5.2.1 Biodiesel

Long-chain alkyl esters, such as biodiesel, are created by chemically or enzymatically transesterifying vegetable or animal fat (Zhang et al. 2020). It has lower emissions and aromatic content, is oxygenated, biodegradable, renewable, less poisonous, and has higher octane numbers and heat content than fossil diesel. It is challenging to utilize directly because of its increased viscosity and poorer ignition quality. A catalyst, methanol or ethanol, and a raw material (oil or fat) are needed for the basic transesterification process (Arumugam and Ponnusam 2017). Idealistically, any plant-based oil can be used; nevertheless, edible oils are not allowed since food must come first, not fuel. Edible oil's high cost makes it unprofitable as well. To create a biodiesel production process that is both economical and efficient, low-cost

Table 5.1 Advanced approaches in biofuel production

Substrate	Alcohol	Microorganism/Enzyme	Advanced approach	Yield	Value added product	References
Rapeseed oil	Methanol	*Candida antarctica*	–	91.1%	–	Watanabe et al. (2007)
Jatropha seed oil	Ethanol	*Candida rugosa*	–	98%	–	Shah and Gupta (2007)
Sunflower oil	Methanol	*Thermomyces lanuginosa*	–	90–97%	–	Dizge et al. (2009)
Waste cooking palm oil	Methanol	*Candida antarctica* B	–	79.1%	–	Halim et al. (2009)
Soybean oil	Methanol	*Rhizomucor miehei*	–	68–95%	–	Guan et al. (2010)
Algal oil	Ethanol	Candida antarctica	–	63.41%	Neutraceutical (EPA C20:5) 50.06%	He et al. (2019)
Babassu oil-derived fatty acid	Ethanol	Rhizomucor miehei	Immobilization on magnetic nanoparticles	81.7%		Moreira et al. et al. (2020)
Oleic acid	Methanol, etha-nol, and butanol	Cold adapted SGNH type lipase (*Halocynthiibacter arcticus*)	Mutagenesis, screening, and immobilization	–	Butyl acetate	Le et al. (2020)
Jatropha curcas oil	Methanol	*Rhizomucor miehei*	Magnetic COF composite and n-hexane as solvent	80%		Zhou et al. (2021)
Olive oil	Methanol	Rhizopus oryzae R1	Enzyme immobilization on chitosan			Helal et al. (2021)

feedstocks like yellow and brown grease, edible oils (like Jatropha oil), and microalgal oils should be utilized. Numerous nations have begun producing fish, castor, soy, palm, and other oils, but the price is too high (Zhang et al. 2020). Because they are more accessible to culture, microbial oils may be a more affordable source.

For this reason, algae and fungi are more suitable sources than bacteria (Zhang et al. 2020). Low-cost waste oils, like sardine oil, are economical choices. Still, contaminants like water and free fatty acids (FFA) make the downstream process challenging using the conventional base catalysis method because FFA can react with the base to form unwanted soap (Arumugam and Ponnusam 2017). A significant amount of methanol is needed for the acid-catalyzed pretreatment of FFA, and recovering acid wastewater is more challenging. Such high FFA oils could be processed by enzymatic catalysis since they produce no soap, simplify downstream processes, and eliminate the requirement for wastewater treatment. A byproduct of high-purity glycerol product could also be obtained. The efficient utilization of waste animal fats in biodiesel synthesis is achieved by applying appropriate enzymes, microwaves, radio-frequency heating, ultrasonic irradiation, and microwaves (Mukhtar et al. 2021).

5.2.2 Biohydrogen

One of the most promising sustainable and eco-friendly energy alternatives is the hydrogen (H_2) industry. Natural gas steam reforming and coal gasification produce 95% H_2 and emit CO_2 (IRENA 2019). Future hydrogen generation methods that are less energy-intensive and cleaner include biological hydrogen production (BHP) and water electrolysis. Direct biophotolysis, indirect biophotolysis, photo-fermentation, and dark fermentation are the four BHP processes. The first two processes need a lot of light to produce H_2 from water. Still, they also need help with the enzymes' sensitivity to oxygen, their poor photochemical efficiency, and the presence of gaseous contaminants. These issues need to be addressed. Although the latter two methods are more cost-effective due to their ability to use a broader range of organic wastes, they have the disadvantage of having lower yields. All that is needed for direct photolysis is sun energy and water. In photosynthetic organisms with PSI and PSII photosystems, ferredoxin generated in light reactions is reoxidized when water is divided. Reduced ferrodoxin produces hydrogen by supplying protons with electrons through a reversible hydrogenase.

$$H_2O \rightarrow PSII \rightarrow PSI \rightarrow Fd \rightarrow Hydrogenase \rightarrow H_2 \qquad (5.1)$$

Thus, cyanobacteria and green algae are promising candidates for green hydrogen production on a large scale. The reductants, ferredoxin/NADH/NADPH, produced in light reactions are needed for CO_2 fixation. However, without CO_2 or O_2, NADPH reduces protons and yields hydrogen gas in a reaction catalyzed by hydrogenase.

[FeFe]-hydrogenase and nitrogenase mediate this reaction in algae and cyanobacteria, respectively (Eq. 5.1). Otherwise, Fd(red) can donate electrons to NAD or NADP, which are subsequently re-oxidized by Group II [NiFe]-hydrogenase in cyanobacteria, also forming H_2. [FeFe]-hydrogenase has 12–14% light-to-hydrogen conversion efficiency, which is considered high (Mona et al. 2020). Research on Chlamydomonas, Chlorella fusca, and Scenedesmus obliquus is being done to produce hydrogen. Anaerobic conditions are necessary for hydrogenase activity, and O_2 uses a negative feedback mechanism to regulate hydrogenase activity. O_2 suppresses enzyme catalysis, gene expression, mRNA stability, and H_2 generation at all molecular levels. As a result, the amount of time that H_2 may be produced is limited, and the only way to produce H_2 continually is to remove O_2 from the system continuously. To produce hydrogen from C. reinhardtii, carbon dioxide (CO_2) limitation, light limitation, and phosphorous (P)/magnesium (Mg) deprivation have all been investigated as ways to accomplish continuous O_2 removal (Ben-Zvi et al. 2019).

Another problem is competition with the Calvin cycle, fueled by electrons exiting photosystems. H_2 generation is merely a tiny electron sink in dark, anoxic environments. By indirect biophotolysis, H_2 generation is geographically or temporally separated from O_2-evolving PS II. Cells with compromised PSII (such as those undergoing sulfur deficiency) or heterocysts may exhibit this. Here, anaerobic fermentation breaks down the starch and glycogen to create reducing equivalents, which, like in biophotolysis, can reduce protons to form H_2. This might happen in bright or dark conditions. Anabaena, Oscillatoria, Nostoc, Calothrix, and other nitrogen-fixing species' heterocysts offer an anaerobic environment for nitrogen fixation:

$$N_2 + 8e^- + 8H^+ + 16ATP \rightarrow 2NH_3 + H_2 + 16ADP + 16Pi \qquad (5.2)$$

A byproduct of this extremely endergonic and irreversible process is hydrogen. NiFe-hydrogenase is required for the synthesis of H_2 in non-nitrogen-fixing taxa such as Gloebacter, Synechococcus, and Synechocystis (Eq. 5.2). Because photofermentative organisms lack PSII and must rely on PSI to obtain ATP, they are unable to emit O_2, which is a clear benefit for producing H_2. Organic or inorganic substrates such as lactate, acetate, sulfite, or ferrous compounds extract the electrons. These substrates lower ferredoxin or NAD through the TCA cycle, and nitrogenase eventually releases H_2. Purple non-sulfur bacteria, such as Rhodobacter capsulatus and Rhodopseudomonas palustris, are the most prevalent photofermentative organisms. Theoretically, photofermentation is the ideal method because it can produce the most H_2 and use waste as a substrate. Still, it has significant limitations due to the need for nitrogen-limiting circumstances and the existence of uptake hydrogenase. The fourth BHP technique, known as "dark fermetation," has the advantages of having a high production rate, not requiring light, and having the capacity to utilize a variety of waste substrates, including lignocellulosic biomass, sewage sludge, and industrial wastewater (Choiron et al. 2020).

5.2.3 Bioethanol

Petrol can be blended with bioethanol in different proportions as a green fuel. Currently, unmodified engines are mixed 5–10% (Singh et al. 2022). Flexible fuel cars are necessary for a further increase in the ethanol proportion. One of the two obstacles preventing the commercialization of bioethanol is its ability to blend with petrol. The cost and accessibility of the raw material or feedstock is the second. In the first generation, Saccharomyces cerevisiae used feedstocks based on sugar and starch for direct and hydrolysis/fermentation, respectively. Starch-containing crops like wheat, barley, corn, cassava, potatoes, and sugarcane are utilized in some nations to produce bioethanol. However, the raw material supply of food crops is inefficient (Bušić et al. 2018). The feedstock for the second generation of bioethanol is lignocellulosic waste, which can be derived from various waste materials, including crop residues, municipal garbage, wood, paper, herbs, and other cellulosic industrial waste. Reducing emissions is an additional benefit of using garbage. However, cellulose has a complicated structure and is more challenging to hydrolyze than starch, which is easier to do. Therefore, one more pretreatment step must be added before hydrolysis and fermentation. The three main categories of this treatment are physical, chemical, and biological, and they differ depending on the raw material utilized. Physical therapies like extrusion, irradiation, pyrolysis, particle size, and crystallinity are reduced; these processes must be optimized with energy requirements in mind. Chemical pretreatment is typically used to solubilize lignin and hemicelluloses and to improve cellulose accessibility. Acids, alkali, organic solvents, and ionic liquids are employed for this purpose; the decision between them is dictated mainly by the impact on the process, including corrosion, the creation of inhibitors, and downstream costs, among other factors. The range of treatments available, matching the variety of feedstocks accessible, includes steam explosion, microwave, ultrasonic, and supercritical fluids. These treatments are designed to increase cellulose accessibility to biological treatment using enzymes. Both bacteria and enzymes can physiologically saccharify cellulose and hemicelluloses. For microbial treatments, brown rot and soft rot fungi are frequently employed. Aspergillus fumigates cultivated on rice straw that had been prepared with NaOH yielded cellulase titers, which were subsequently utilized to ferment the same substrate to create ethanol. Cellulase is the main ingredient for enzymes, while additional enzymes can be used to increase output. Pretreatments must be combined and optimized based on the feedstock's recalcitrance and source. Numerous lab research studies have been conducted on various waste lignocellulosic feedstocks. A tropical plant called vetiver grass is utilized to conserve water and soil. Under low nutritional conditions, it generates a high biomass. This grass was used by Subsamran et al. (2019) to produce cellulase and bioethanol (maximum 5.85 gL1) by combining chemical and enzymatic processing. Following soda and enzyme-catalyzed bioextrusion, barley straw yields reasonable amounts of ethanol (38 gL1) (Duque et al. 2020). Following saccharification, soluble sugars are converted to ethanol through fermentation by microorganisms, most frequently the yeast Saccharomyces cerevisiae. Although it can only ferment glucose, fructose, and sucrose, the bacteria Zymomonas mobilis is also a promising ethanol fermentor since it produces more and is more productive than S. cerevisiae (Lin and Tanaka 2006). Another alternative is to engineer strains of E. Coli that ferment a wide range of carbohydrates, including all lignocellulosic sugars (hexoses and pentoses). Although productivity and yields are poor, Klebsiella oxytoca and Pichia stipitis yeast can also make ethanol from pentose carbohydrates. The organism's tolerance to greater amounts of ethanol is a significant determining factor. In addition to using pentoses, primarily xylose and arabinose, lignocellulosic substrates also require utilization. Pentose catabolizing pathways have thus been added to the ethanol-producing bacteria S. cerevisiae and Z. mobilis.

Algae provide the ethanol used in the third generation. Compared to lignocellulosic, algal biomass has two main advantages as a feedstock: first, it can be grown more quickly, and second, it can absorb CO_2 more quickly. Additionally, leftover biomass has fertilizer potential. Algal feedstocks are devoid of lignin and comprise glucose and galactose. Cellulase is needed for the pretreatment, similar to lignocellulosic. Amylases are required for the hydrolysis of substrates from the first generation. Modern industrial processes use amylases and related enzymes such as pullulanase, glucosidase, and others for hydrolysis, which is a simple process. Commercial enzyme preparations work well on various starchy substrates and can be used anywhere. Shanavas et al. (2011) employed SpezymeXtra and Stargen 001 on cassava starch and achieved large yields (558 g ethanol/kg starch) and excellent efficiency (98%) in 48 h. Fascinatingly, rye and wheat seeds grown in situ using in situ enzymes generated during germination had high autoamylolytical values (90%) compared to commercial enzymes by Kadar et al. (2011).

5.2.4   Biogas

Biogas is a valuable renewable energy source formed from organic materials that decompose naturally and carry secondary Energy. It often consists of a mixture of gases, including siloxanes, moisture, carbon dioxide (CO_2), methane (CH_4), and very little hydrogen sulfide (H_2S). Production sources state that the composition of biofuel varies, with the main constituents being CH_4 (45–70%), CO_2 (30–40%), and N_2 gas (1–15%). With the development and improvement of affordable bioprocesses, the production of biogas and biomass-based Energy such as biohydrogen and bioethanol would be a great alternative to Energy obtained from fossil fuels, providing more opportunities to people who live in rural areas of the world. A significant amount of Energy is released during the oxidation or combustion of H_2, CH_4, or CO in conjunction with oxygen, and this Energy is used for cooking or heating (Singh et al. 2022). According to Awe et al. (2017), biogas is a starting material or fuel source for synthesizing chemicals, hydrogen, and synthetic gases. Renewable heat or electricity has been produced using stored biomass energy. Biopower has been produced by gasification, or the combustion of dry biomass or biogas (methane), under the supervision of anaerobic bacteria (Awe et al. 2017). Direct burning of small, compressed wood masses, leftover wood after cutting, and other dry biomass components can create thermal power for industrial and domestic heating and cooking at high temperatures and pressures as long as oxygen is available. Cost-effective process improvements result in a decrease in greenhouse gas production. This process's main selling points are its versatility and ease of application with nearly any organic substrate. In anaerobic digestion, microorganisms break down organic matter in several processes without oxygen. Due to this process, methane, or marsh gas, naturally forms at the bottom of ponds and marshes. In addition to providing all the benefits of green fuel, biogas production is also the most profitable.

The green economy is favored by the reduction of trash's volume and odor, the value addition of waste as fertilizer, and the production of power and heat (Fig. 5.1). In actuality, research on biogas has advanced to the bioeconomy approach where systems that maximize waste potential by producing different products from a single processing of garbage based on area are being investigated. An excellent example may be found in a publication by Malayil and Chanakya (2016), where they discuss using biogas plant sludge as a nutrition for growing mushrooms. Paddy straw was pretreated with leachate from mushroom growing to produce biogas. While the production of biogas increased by 29%, the yields of mushrooms climbed by 44%. Numerous microbial communities must cooperate either sequentially or

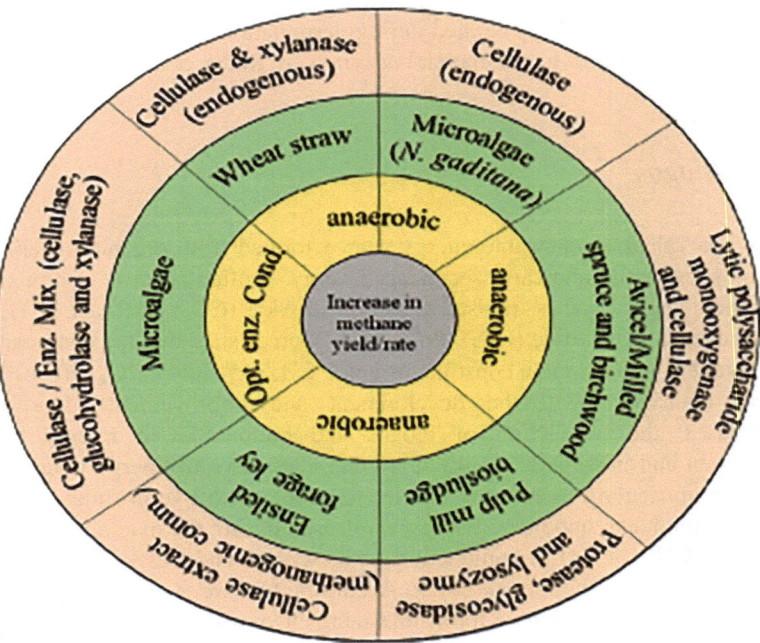

Fig. 5.1 Integrated approach for methane production

simultaneously to produce biogas. The decomposition process involves multiple stages, each requiring a distinct group of microorganisms with unique needs in terms of nutrition and habitat. Figure 5.1 illustrates the breakdown of giant organic molecules into carbon dioxide and methane in four steps.

5.3 Enzymes in Biofuel Production

Chemical and biological processes are the two main categories into which biofuel production methods may be divided. A biochemical process uses biological catalysts, or enzymes, to transform raw materials into fuel, whereas a chemical process uses heat and chemical catalysts. According to Al-Zuhair et al. (2011), using enzymes can address the drawbacks of using traditional chemical catalysts, such as their energy intensity and environmental friendliness.

5.3.1 Amylases

The initial stage of producing biofuels involves treating starch molecules at α-1,4 links with α-amylases to hydrolyze them and create short chains of dextrins. Glucoamylases hydrolyze these short chains to create maltose, glucose, and isomaltose, which are subsequently readily fermented to make butanol or ethanol. According to Ajita et al. (2014), Penicillium and Aspergillus are two fungus sources, and Bacillus is the most commonly used bacterial species for amylase production.

5.3.2 Lignocellulosic Biomass

The raw material used in the second generation of biofuel production was lignocellulosic biomass. It has a lot of promise as a raw material for producing biomass. It comprises cellulose, hemicellulose, and lignin in a 3:3:4 ratio. These are first pre-treated, and after that, saccharification enzymes break them down enzymatically into simpler sugars, which are subsequently fermented to produce the required chemical result, like bioethanol.

5.3.2.1 Cellulase

Cellulose is a linear polysaccharide comprising glucose units that repeat and are joined by β-1,4-glycosidic bonds. Cellulose consists of 15–45 repeating units of glucan. The arrangement of glucan units into fibrils results in microfibrils, which give cellulose a structure that alternates between crystalline and amorphous. This structure protects it from deterioration by increasing the surface area. One common industrial enzyme is cellulase. The enzyme cellulase breaks down Complex cellulose sugar molecules into simpler glucose sugars. It is known that several fungal species make cellulase.

Cellulose is a linear polysaccharide comprising glucose units that repeat and are joined by β-1,4-glycosidic bonds. Cellulose consists of 15–45 repeating units of glucan. The arrangement of glucan units into fibrils results in microfibrils, which give cellulose a structure that alternates between crystalline and amorphous. This structure protects it from deterioration by increasing the surface area. One common industrial enzyme is cellulase. The enzyme cellulase breaks down Complex cellulose sugar molecules into simpler glucose sugars. It is known that several fungal species make cellulase. Cellulase is produced commercially by microorganisms like Trichoderma reesei and Aspergillus niger (Srivastava et al. 2020). Three enzymes make up the cellulase enzyme system:

(a) Endoglucanases
(b) Exoglucanases
(c) β-glucosidases

Cellulase is a difficult-to-reuse enzyme with limited enzyme stability. Several strategies increase enzyme stability, including protein engineering and immobilization. One crucial stage in the manufacturing of biofuel that determines cost is the conversion of cellulose to sugar.

5.3.2.1.1 Endoglucanase

Endoglucanase, called carboxymethyl cellulase (CMCase), breaks down internal links in amorphous cellulose chains, causing a nick to form randomly. This process converts lignocelluloses into monomeric sugars (Narra et al. 2014). They are regarded as affordable because of their increased specificity and reduced energy intake (Daniel et al. 2012). Bacterial strains like Streptomyces sp. and Bacillus sp. combine fungus strains like Aspergillus and Trichoderma to manufacture commercial endoglucanase. They are members of the family of glycosyl hydrolases. The glucoside hydrolases known as endoglucanases, or GH5, GH6, GH7, GH9, GH12, GH44, GH45, GH48, GH51, GH74, and GH124, have been divided into families. Endoglucanases hydrolyze cellulose by two different mechanisms: retention and inversion.

5.3.2.1.2 Cellobiohydrolase

Cellobiohydrolase (CBH), or exoglucanase, functions with β-glucosidase and endoglucanase to hydrolyze cellulose. One type of exocellulase enzyme is cellobiohydrolase. By cleaving 1,4-β-Dglycosidic linkages from cellulose terminals, it breaks down cellulose molecules. CBH1 and CBH2 are the two different forms of cellobiohydrolases. CBH1 and CBH2 cleave the cellobiose by moving along cellulose's reducing and non-reducing ends, respectively. The rate at which cellulosic breakdown occurs overall is thought to be limited by the hydrolysis of cellulose into cellobiose. This includes the exo-endo synergism between endoglucanase and cellobiohydrolase and the exo-exo synergism between CBH1 and CBH2. This makes it easier to remove the remaining cellobiose molecule residue from cellulose's reducing and non-reducing ends (Liu et al. 2011). Fungal species like Trichoderma reesei, Penicillium sp., and Polyporus sp. (Fang and Xia 2015) and bacterial strains like Flavobacterium sp. and Paenibacillus sp. (Islam and Roy 2019) are used in the commercial production of CBH. The substrate is closer to the glycosyl hydrolase when CBH is present. This modifies the surface crystals of cellulose and improves accessibility (Sánchez et al. 2003).

5.3.2.1.3 β-Glucosidases

Exoglucanase operates on these exposed chains to release the cellobiose and glucose. In contrast, endoglucanase exposes the non-reducing and reducing ends of cellulose polysaccharide chains by creating nicks. β-glucosidases (BGL) are involved in the saccharification of cellulose to simpler sugars, and they are the most critical and final step in this process. BGL carries out the last step of the process, transforming the cellobiose that exoglucanase produces into glucose (Horn et al. 2012). While cellobiose inhibits exoglucanase and endoglucanase, it keeps the overall rate of cellulase hydrolysis constant. BGL is a rate-limiting enzyme due to glucose's feedback inhibition of the enzyme. Therefore, to create commercially viable procedures for the generation of biofuel, an effective BGL that is tolerant of high glucose concentrations is required (De Andrades et al. 2019). Microorganisms well-documented for producing BGL include Aspergillus sp., Penicillium sp., and certain Bacillus species. While fungi, mammals, plants, and bacteria may all manufacture BGL, bacteria are the most recommended option for commercial use because of their quick growth, simplicity of handling, and engineering capabilities.

The biofuel sector requires a higher level of BGL production because of its lower production and higher cost. The rate of hydrolysis is accelerated by BGL immobilization (Muhammad et al. 2016). It is anticipated that this method will produce biofuel faster and cheaper.

5.3.2.2 Xylanases

Hexose and pentose sugar units make up the branching heteropolymer known as hemicellulose. According to Beg et al. (2001), xylan is the most common hemicellulose and contains β-Dxylopyranosyl residues with β-1,4-glycosidic connections. The cell wall components include cellulose, xylan, and lignin, which are joined by covalent and noncovalent connections. It is found where cellulose and lignin converge, an essential location for preserving the integrity of the cell wall. The glycoside hydrolase (GH) family of enzymes includes xylanase, which breaks down xylan. The combined action of several xylanases fully hydrolyzes the xylan backbone.

5.3.2.3 Laccase

Lignin is the second most prevalent naturally occurring polymer, behind cellulose. It is the only aromatic feedstock that is renewable. It significantly contributes to the structure of plant cell walls. It is a heteropolymer with a variety of structures. According to Bugg et al. (2021), it is a complex aromatic polymer comprising phenylpropanoid aryl-C3 units joined by CC and CO links. Because of its chemically robust structure, lignin removal from lignocellulosic biomass presents a significant hurdle in biofuel generation. Thus, the lignin must be pre-treated. The three distinct enzymes found in ligninase—manganese peroxidase, Laccase, and lignin peroxidase—are necessary for the full breakdown of lignin (Wang et al. 2017).

All three enzymes' lignin breakdown is thoroughly researched about Laccase. It is a member of the multicopper oxidase (MCO) family of copper-containing oxidases. Because of their enhanced system for expressing the laccase gene family involved in lignin degradation, white-rot fungi have evolved naturally to express Laccase in large quantities. Several fungal genera produce laccase, including deuteromycetes, ascomycetes, and basidiomycetes. Marinomonas mediterranea, Streptomyces lavendulae, Bacillus subtilis, Geobacillus thermocatenulatus, and Aquisalibacillus elongatus are a few of the bacteria that produce Laccase (Srivastava et al. 2020). The polysaccharides in lignocellulosic biomass must be fully hydrolyzed to produce ethanol. This can be accomplished by pretreating the biomass. One possible agent for the lignin removal process is Laccase.

5.3.2.4 Pectinases

Pectin is a polymer linked to other polymers and carbohydrates by a backbone chain of rhamnogalacturonan molecules. According to Munarin et al. (2013), it is a linear polymer of D-α-(1,4) anhydrous-galacturonic acid. Peptidases break down pectin. Next to the free carboxyl group, glycosidic bands are hydrolyzed by polygalacturonases (PG). Pectinase lyase (PL) hydrolyzes using the β elimination mechanism. The pectin chain is divided at random by both enzymes. PG is produced by the fungi Aspergillus sp. and Penicillium sp. Pectinase is known to be made by Bacillus species. Pectinases are involved in the conversion of plant cell walls to bioethanol during the biofuel production process. Pectinases aid in the breakdown of enzymes when lignin levels are decreased (Srivastava et al. 2020). New extraction methods are needed to lower the price of pectinases and facilitate the cost-effective manufacture of biofuel.

5.3.2.5 Proteases

Protease is the enzyme that uses proteolysis to break down the protein in the biomass used to produce biofuel. These could be endopeptidases, which dissolve internal bonds in polypeptide chains, or exopeptidases, which extract amino acids from carbon and nitrogen termini (Domsalla and Melzig 2008). Numerous fungi, yeasts, and bacteria, including Penicillium roqueforti, Bacillus subtilis, and Aspergillus oryzae, are among those that produce proteases. Proteases increase the effectiveness of microorganisms in biofuel production by releasing nitrogen and increasing the amount of alcohols released from branched-chain 2-keto acid for biofuel conversion (Huo et al. 2011).

5.4 Genomic Engineering/Genetic Engineering: Bridging the Gap Between Lab to Commercial

Today's widespread fuel use has a significant negative impact on the environment. Ethanol is acknowledged as one of the most important renewable fuels among all biofuels. It is a novel energy source with a complex blending process but an excellent antiknock value and less adverse environmental effect. Energy-rich crops like maize, corn, sugarcane, lignocellulosic biomass, and algae biomass are needed to produce ethanol, depending on the generation. Producing ethanol from lignocellulosic biomass is intricate, requiring hydrolysis, proper pretreatment, and fermentative bacteria to break it down. Due to growing environmental concerns over the past 10 years, the energy sector has focused on creating bioethanol using affordable technology. Many research institutes, commercial companies, and industries place a high premium on genetic engineering tools for ethanol production. In this field of study, microorganisms are metabolically and genetically modified to yield the required amount of ethanol. The utilization of microorganisms in the catalysis of biomass conversion into biofuels is contingent upon the organism's capacity to produce biofuels at a faster and more cost-effective rate on an industrialized scale. Microbes that have been metabolically modified to convert renewable carbon sources into desired fuel products are being researched as the best substitute for high volumetric yield and potency.

Many microorganisms, which are environmental isolates, possess unique metabolic pathways that convert biomass into a product comparable to biofuels. Specific strains of microorganisms that maximize the pace of biochemical reactions for energy transfer and biofuel generation have been engineered to facilitate the conversion of biomass materials into different types of biofuel. To hydrolyze biomass enzymes, protein transporters, or passageways for carbon uptake, CO_2 fixation, and photosynthetic processes having varied metabolic roles for the production of biofuel, microalgal species are, therefore, a marvelous platform (Singh et al. 2022). Different microbial strains are employed to increase the production of sustainable biofuels due to the advancement of synthetic biology and engineering techniques (Lee and Lee 2013).

For instance, using the Hinshelwood model, it has been investigated how varying reducing sugar concentrations, or 85–156 g L1, affect the kinetic growth parameters of ethanol production using sweet sorghum as the substrate and the CICC 1308 strain of S. cerevisiae (Jin et al. 2020). Random chemical mutagenesis, UV radiation exposure, and genome shuffling have all been used to modify microbial systems with precise knowledge of cell tolerance capacity and cell defense adaption during the production of alcohol fuel. The technique of producing Brazilian ethanol has been claimed to involve the deletion of the GPD2 gene, which codes for glycerol 3-phosphate dehydrogenase, and the overexpression of the GLT1 gene, which codes for the glutamate synthase enzyme, with the aid of the yeast S. cerevisiae. Yeast may be cultivated at high temperatures and cell density per unit volume (Abreu-Cavalheiro and Monteiro 2013). Microbial fermentation has also been touted as a less expensive and more efficient method of producing sustainable biodiesel for various biotechnology industry start-ups and research institutes.

5.5   Conclusions and Future Perspectives

Because they are a sustainable and environmentally beneficial substitute for quickly running out of fossil fuels, biofuels have gained popularity. But as of right now, biofuels are not profitable. While biological synthesis suffers from poor effective activity and other inefficiencies, its chemical synthesis requires a lot of harsh chemicals and is Energy inefficient. It is still difficult to digest lignocellulosic biomass enzymatically. Using metagenomics-based techniques, new enzymes from uncharted settings can be identified. Even though many enzymes have been screened and determined, more studies are required to standardize and scale up the procedure. Future opportunities for the generation of biofuels are increased by knowing how different enzymes from termite and fungal biomes break down lignin. A wider variety of substitute hosts with high-quality foreign gene expression from metagenomic origins must be developed. Several methods, such as metabolic engineering using genetic engineering to modify pathways and optimize the synthesis of desired metabolites, have been explored and proven effective in the lab. However, only a few species have been the subject of this research. One of the main obstacles to efficient metabolic design is the need for more understanding of metabolic pathways. However, as more genomes are sequenced, improved bioinformatics tools are developed, and metabolic pathways are studied, better systems for efficiently synthesizing biofuels will be designed. To make biofuel manufacturing economically feasible, more research is required.

References

Abreu-Cavalheiro A, Monteiro G (2013) Solving ethanol production problems with genetically modified yeast strains. Braz J Microbiol 44:665–671

Ajita S, Pandurangappa T, Murthy K (2014) α-Amylase production and applications: a review. J Appl Environ Microbiol 2(4):166–175. https://doi.org/10.12691/jaem-2-4-10

Al-Zuhair S, Ramachandran K, Farid M, Aroua M, Vadlani P, Ramakrishnan S, Gardossi L (2011) Enzymes in biofuels production. Enzym Res (1):658263. https://doi.org/10.4061/2011/658263

Arumugam A, Ponnusam V (2017) Production of biodiesel by enzymatic transesterification of waste sardine oil and evaluation of its engine performance. Heliyon 3:1–486

Awe OW, Zhao Y, Nzihou A, Minh DP, Lyczko N (2017) A review of biogas utilisation, purification and upgrading technologies. Waste Biomass Valoriz 8:267–283

Beg QK, Kapoor M, Mahajan L, Hoondal GS (2001) Microbial xylanases and their industrial applications: a review. Appl Microbiol Biotechnol 56:326–338

Ben-Zvi O, Dafni E, Feldman Y, Yacoby I (2019) Re-routing photosynthetic energy for continuous hydrogen production in vivo. Biotechnol Biofuels 12:1–3

Brunerová A, Maľaťák J, Müller M, Valášek P, Roubík H (2017) Tropical waste biomass potential for solid biofuels production. ABSGGL 15:359–368

Bugg TD, Williamson JJ, Alberti F (2021) Microbial hosts for metabolic engineering of lignin bioconversion to renewable chemicals. Renew Sust Energ Rev 152:111674

Bušić A, Mardetko N, Kundas S, Morzak G, Belskaya H, Ivančić M, Šantek B, Komes D, Novak S, Šantek B (2018) Bioethanol production from renewable raw materials and its separation and purification: a review. Food Technol Biotechnol 56:289–311

Choiron M, Tojo S, Chosa T (2020) Biohydrogen production improvement using hot compressed water pretreatment on sake brewery waste. Int J Hydrog Energy 45:17220–17232

Daniel KM, Piotr OP, Simmons BA, Blanch HW (2012) The challenge of enzyme cost in producing Lignocellulosic biofuels. Biotechnol Bioeng 109:1083–1087

De Andrades D, Graebin NG, Ayub MAZ, Fernandez-Lafuente R, Rodrigues RC (2019) Physico-chemical properties, kinetic parameters, and glucose inhibition of several beta-glucosidases for industrial applications. Process Biochem 78:008–090. https://doi.org/10.1016/j.procbio.2019.01.008

Dizge N, Aydiner C, Imer DY, Bayramoglu M, Tanriseven A, Keskinler B (2009) Biodiesel production from sunflower, soybean and waste cooking oils by transesterification using lipase immolbilized onto a novel microporous polymer, vol 100. Bioresour Technol, pp 1983–1991

Domsalla A, Melzig M (2008) Occurrence and properties of proteases in plant latices. Planta Med 74(07):699–711

Duque A, Doménech P, Álvarez C, Ballesteros M, Manzanares P (2020) Study of the bioprocess conditions to produce bioethanol from barley straw pre-treated by combined soda and enzyme-catalyzed extrusion. Renew Energy 158:263–270

Fang H, Xia L (2015) Heterologous expression and production of *Trichoderma reesei* cellobiohydrolase II in *Pichia pastoris* and the application in the enzymatic hydrolysis of corn Stover and rice straw. Biomass Bioenergy 78:99–109

Guan F, Peng P, Wang G, Yin T, Peng Q, Huang J, Guan G, Li Y (2010) Combination of two lipases more efficiently catalyzes methanolysis of soybean oil for biodiesel production in aqueous medium. Process Biochem 45(10):1667–1682

Halim SFA, Kamaruddin AH, Fernando WJN (2009) Continuous biosynthesis of biodiesel from waste cooking palm oil in a packed bed reactor: optimization using response surface methodology (RSM) and mass transfer studies. Bioresour Technol 100:710–716

He Y, Wang X, Wei H, Zhang J, Chen B, Chen F (2019) Direct enzymatic ethanolysis of potential Nannochloropsis biomass for co-production of sustainable biodiesel and nutraceutical eicosapentaenoic acid. Biotechnol Biofuels 12:1–5

Helal SE, Abdelhady HM, Abou-Taleb KA, Hassan MG, Amer MM (2021) Lipase from Rhizopus oryzae R1: in-depth characterization, immobilization, and evaluation in biodiesel production. J Genet Eng Biotechnol 19:1–3

Horn SJ, Vaaje-Kolstad G, Westereng B, Eijsink VGH (2012) Novel enzymes for the degradation of cellulose. Biotechnol Biofuels 5:45. https://doi.org/10.1186/1754-6834-5-45

Huo YX, Cho KM, Rivera JGL, Monte E, Shen CR, Yan Y et al (2011) Conversion of proteins into biofuels by engineering nitrogen flux. Nat Biotechnol 29:346–351

IRENA (2019) Renewable energy statistics. "International renewable energy agency." Renewable Energy Target Setting, Abu Dhabi

Islam F, Roy N (2019) Isolation and characterization of cellulase-producing bacteria from sugar industry waste. Am J Biosci 7(1):16–24

Jin X, Song J, Liu GQ (2020) Bioethanol production from rice straw through an enzymatic route mediated by enzymes developed in-house from Aspergillus fumigates. Energy J 190:116–395

Kádár Z, Christensen AD, Thomsen MH, Bjerre AB (2011) Bioethanol production by inherent enzymes from rye and wheat with addition of organic farming cheese whey. FUELAC 90:3323–3329

Kumar A, Sharma S, Pandey LM, Chandra P (2018) Nanoengineered material based biosensing electrodes for enzymatic biofuel cells applications. Mater Sci Energy Technol 1(1):38–48

Le LT, Yoo W, Jeon S, Lee C, Kim KK, Lee JH, Kim TD (2020) Biodiesel and flavor compound production using a novel promiscuous cold-adapted SGNH-type lipase (Ha SGNH1) from the psychrophilic bacterium Halocynthiibacter arcticus. Biotechnol Biofuels 13:1–3

Lee SJ, Lee DW (2013) Design and development of synthetic microbial platform cells for bioenergy. Front Microbial 19:4–92

Lin Y, Tanaka S (2006) Ethanol fermentation from biomass resources: current state and prospects. Appl Microbiol Biotechnol 69:627–642

Liu N, Yan X, Zhang M et al (2011) Microbiome of fungus-growing termites: a new reservoir for Lignocellulase genes. Appl Environ Microbiol 77(1):48–56

Malayil S, Chanakya HN (2016) Fungal enzyme cocktail treatment of biomass for higher biogas production from leaf litter. Procedia Environ Sci 35:826–832

Malode SJ, Gaddi SAM, Kamble PJ, Nalwad AA, Muddapur UM, Shetti NP (2022) Recent evolutionary trends in the production of biofuels. Mater Sci Energy Technol 5:262–267

Mona S, Smita S, Kumar V, Parveen K, Saini N, Bansal D (2020) Pugazhendhi, green technology for sustainable biohydrogen production (waste to energy): a review. Sci Total Environ 728:138–481

Moreira KD, de Oliveira AL, Júnior LS, Monteiro RR, da Rocha TN, Menezes FL, Fechine LM, Denardin JC, Michea S, Freire RM, Fechine P (2020) Lipase from Rhizomucor miehei immobilized on magnetic nanoparticles: performance in fatty acid ethylester (FAEE) optimized production by the Taguchi method. Front Bioeng Biotechnol 8:693

Muhammad J, Buthe A, Rashid M, Wang P (2016) Cost-efficient entrapment of b-glucosidase in nanoscale latex and silicone polymeric thin films for use as stable biocatalysts. Food Chem 190:1078–1085. https://doi.org/10.1016/j.foodchem.2015.06.040

Mukhtar A, Saqib S, Mubashir M, Ullah S, Inayat A, Mahmood A, Ibrahim M, Show PL (2021) Mitigation of CO2 emissions by transforming to biofuels: optimization of biofuels production processes. Renew Sust Energ Rev 150:111487

Munarin F, Tanzi MC, Petrini P (2013) Corrigendum to 'Advances in biomedical applications of pectin gels'. Int J Biol Macromol 55:307

Narra M, Dixit G, Divecha J, Kumar Adepu K, Madamvar D, Amita S (2014) Production, purification and characterization of a novel GH 12 family endoglucanase from aspergillus terreus and its application in enzymatic degradation of delignified rice straw. Int Biodeterior Biodegrad 88:150–161. https://doi.org/10.1016/j.ibiod.2013.12.016

Sánchez MM, Pastor FJ, Diaz P (2003) Exo-mode of action of cellobiohydrolase Cel48C from Paenibacillus sp. BP-23: a unique type of cellulase among Bacillales. Eur J Biochem 270(13):2913–2919

Selvarajoo A, Wong YL, Khoo KS, Chen WH, Show PL (2022) Biochar production via pyrolysis of citrus peel fruit waste as a potential usage as solid biofuel. Chemosphere 294:133671

Shah S, Gupta MN (2007) Lipase catalyzed preparation of biodiesel from Jatropha oil in a solvent free system. Process Biochem 42:409–414

Shanavas S, Padmaja G, Moorthy SN, Sajeev MS, Sheriff JT (2011) Process optimization for bioethanol production from cassava starch using novel ecofriendly enzymes. Biomass Bioenergy 35:901–909

Singh AR, Singh SK, Jain S (2022) A review on bioenergy and biofuel production. Mater Today Proc 49:510–516

Srivastava N, Mishra PK, Upadhyay SN (2020) Microbial cellulase production. In: Industrial enzymes for biofuels production, pp 19–35. https://doi.org/10.1016/B978-0-12-821010-9.00002-4

Subsamran K, Mahakhan P, Vichitphan K, Vichitphan S, Sawaengkaew J (2019) Potential use of vetiver grass for cellulolytic enzyme production and bioethanol production. Biocatal Agric Biotechnol 17:261–268

Wang Z, Liu J, Ning Y, Liao X, Jia Y (2017) Eichhornia crassipes: agro-waster for a novel thermostable laccase production by Pycnoporus sanguineus SYBC-L1. J Biosci Bioeng 123(2):163–169. https://doi.org/10.1016/j.jbiosc.2016.09.005

Watanabe Y, Pinsirodom P, Nagao T, Yamauchi A, Kobayashi T, Nishida Y, Takagi Y, Shimada Y (2007) Conversion of acid oil by-produced in vegetable oil refining to biodiesel fuel by immobilized Candida Antarctica lipase. J Mol Catal B Enzym 44(3–4):99

Zhang P, Chen X, Leng Y, Dong Y, Jiang P, Fan M (2020) Biodiesel production from palm oil and methanol via zeolite derived catalyst as a phase boundary catalyst: an optimization study by using response surface methodology. Fuel 272:117680

Zhou ZW, Cai CX, Xing X, Li J, Hu ZE, Xie ZB, Wang N, Yu XQ (2021) Magnetic COFs as satisfactory support for lipase immobilization and recovery to effectively achieve the production of biodiesel by maintenance of enzyme activity. Biotechnol Biofuels 14:1–2

Chapter 6
Recovery of Enzymes from Food Wastes

Abstract The Food and Agricultural Organization (FAO) estimates that around 1.3 billion tonnes, or one-third, of the food produced worldwide for human consumption is lost during the supply chain. Food waste is now landfilled or burned with other combustible municipal garbage in several countries to recover energy potentially. However, the pressures from the economy and the environment are mounting for these two choices. Food waste can be turned into valuable goods (such as bio-products like methane, hydrogen, ethanol, enzymes, organic acids, chemicals, and fuels) by various fermentation methods because of its organic and nutrient-rich nature. This conversion method may be more profitable than turning food waste into animal feed or fuel for transportation. As a result, there is increased interest in value-added bioproducts made from food waste, including fuels, chemicals, enzymes, hydrogen, ethanol, and organic acids. Thus, this review aims to present information on food waste, focusing on Asia-Pacific nations and cutting-edge methods for processing food waste to create enzymes.

Keywords Amylase · Protease · Lipase · Lignocellulosic enzymes · Pectinolytic enzymes

6.1 Introduction

Food waste (F.W.) is the organic waste generated in restaurants, cafeterias, commercial and residential kitchens, and food processing facilities. Around the world, it makes up a considerable amount of municipal solid trash (Lundqvist et al. 2008). According to FAO, nearly 1.3 billion tonnes of food, including fresh fruits, vegetables, meat, bread, and dairy products, are wasted within the food supply chain (2012). Population growth and economic expansion, especially in Asian countries, are driving up the F.W. According to Melikoglu et al. (2013), the yearly volume of urban F.W. in Asian countries will increase from 278 to 416 million tonnes between 2005 and 2025. China (82.8 million tonnes (M.T.) produced the most in a single year), followed by Indonesia (30.9 MT), Japan (16.4 MT), the Philippines (12 M.T.), and Vietnam (11.5 MT). However, with 280 kg generated annually, New Zealand and Australia produced the most fisheries waste (F.W.) per person. In Southeast

Asia, the only country with lower F.W. production per capita was Cambodia (173 kg).

Food waste seems to be more common in high-income states; however, although China has the highest absolute amount of food waste, its waste production per capita is the lowest (61 kg/year), while Singapore and Hong Kong have the highest waste production per capita (168 kg/year and 120 kg/year, respectively) (Lin et al. 2013). Food wastes have been disposed of as animal feed, disposed of in landfills, burned, composted, and anaerobically digested. F.W. is still disposed of in landfills, dumpsites, and other domestic garbage in many Asian nations. Regretfully, landfills have a finite capacity and cannot accommodate growing volumes of F.W. (Ngoc and Schnitzer 2009). F.W. is typically burned with other combustible municipal solid wastes to provide energy or heat to minimize its volume, especially in Singapore and Japan.

Regarding greenhouse gas emissions and total energy recovery, burning F.W. is preferable to landfilling (Othman et al. 2013). However, because of the high initial and ongoing expenditures, incineration could not be a viable option for most low-income nations (Ngoc and Schnitzer 2009). Furthermore, it should be mentioned that burning F.W. has the potential to pollute the air (Takata et al. 2012). Composting is another method for managing biodegradable firewood, yielding a beneficial fertilizer and soil conditioner (Gajalakshmi and Abbasi 2008). Compared to other treatment techniques, composting has a high economic efficiency and a comparatively low environmental impact. However, F.W.'s high moisture content will cause a significant leachate leak. Compost costs more than commercial fertilizers, and there is only a tiny market for compost made from F.W.

6.2 Enzymes from Plant Food Processing Wastes

6.2.1 Peroxidases (PODS)

PODs are frequently found in plant debris rejected after processing edible plant tissues. This material can be considered an inexpensive source of PODs, which may have promising futures in various environmental and food treatment applications. However, only a tiny number of PODs from various residual sources have undergone extensive testing for use in different processes, even though PODs from several residual sources, such as spring cabbage (Belcarz et al. 2008), broccoli (Duarte-Vázquez et al. 2007), asparagus (Jaramillo-Carmona et al. 2013), lentil stubbles (Hidalgo-Cuadrado et al. 2012), and wheat bran, have been studied concerning their biochemical properties. The most extensively researched POD is the enzyme found in horseradish (Armoracia rusticana) roots. Although not strictly speaking a by-product of plant processing, horseradish roots are the primary source of plant POD. As members of the class III peroxidases (classical secretory plant peroxidases), horseradish POD (HRP), peanut, soybean (SBP), turnip, tomato, and barley PODs catalyze the oxidation of a wide range of electron donor substrates by H_2O_2.

HRP catalyzes the oxidative coupling of phenolic compounds utilizing H_2O_2 as the oxidizing agent in heme peroxidases (EC 1.11.1.7), including a ferriprotoporphyrin IX prosthetic group at the active site. The enzyme is reduced in two consecutive one-electron transfer steps from reducing substrates, usually a phenol derivative (Derat and Shaik 2006), in a three-step cyclic process that begins with oxidation by H_2O_2. The application of HRP for bioremediation has received the most attention. Early research demonstrating HRP's capacity to act on aromatic amines and remove them from aqueous media has highlighted the substance's value as a detoxifying agent with potential in bioremediation (Klibanov and Morris 1981). Subsequent research to this groundbreaking study demonstrated the effectiveness of HRP in oxidizing a broad spectrum of industrial organic pollutants, including bisphenol A, dyes, phenols, cresols, chlorophenols, estrogens, and polychlorinated biphenyls. Most of these investigations showed that HRP was incredibly effective at eliminating contaminants through the generation of radicals, polymerization, and sedimentation of the resulting insoluble polymers—often with the help of an additive like polyethylene glycol (PEG), Tweens, etc. The enzyme's high effectiveness was typically observed in pH values between 5 and 7, regardless of whether it was immobilized or free (soluble). However, treatments could also result in pH values between 2.5 and 9, primarily based on the type of additive and the target molecule.

Similarly, 25 °C was the ideal temperature for the investigated therapies. This temperature is reasonably appropriate for the enzyme to exhibit high activity and is nearly identical to what is experienced in working settings. Even though there are many reports on the enzyme efficiency tests conducted in aqueous solutions, the first research showed the effectiveness of free enzymes by treating bleaching plant effluent, which resulted in the removal of more than 60% of industrial dye (Paice and Jurasek 1984). Dec and Bollag (1994) demonstrated the effectiveness of using minced horseradish as a crude source of POD, leading to the total elimination of 2,4-dichlorophenol from an industrial effluent. Similarly, multiple studies have demonstrated that POD from SBP is most successful at eliminating various resistant pollutants between pH values of 5.5 and 9.

It is interesting to note that in several instances, whole soybean seed hulls (Geng et al. 2004) or crude enzyme extracts (Al-Ansari et al. 2009), a regularly generated soybean processing waste, were used in place of purified enzymes, yielding removal yields of more than 95% of common toxicants, such as phenol and chlorophenols. A comparable result with chopped potatoes has also been shown (Dec and Bollag 1994). This discovery is crucial since enzyme purification procedures would be far more expensive. A recent comparison analysis shows that SBP has the best thermo-stability and is a better biocatalyst for removing 4-chlorophenol than artichoke POD and HRP. PODs from a few other sources seemed like viable bioremediation methods; in certain instances, the ability to remediate industrial wastewater well-substantiated this ability. It has been demonstrated that a crude extract from onion solid wastes can reduce the polyphenol content of olive mill wastewater (OMW) by over 50% (Barakat et al. 2010). Industrial effluents treated with minced potato (Dec and Bollag 1994) and immobilized turnip POD likewise showed reductions of 2,4-dichlorophenol and other phenolics of more than 88%. When added to model effluents, crude enzyme preparations from bitter gourd (Satar and Husain 2009), turnip (Duarte-Vázquez et al. 2002), and artichoke (López-Molina et al. 2003) showed varying degrees of efficacy, up to 96% elimination.

Although there is a wealth of research on using PODs derived from waste plants for bioremediation, the data also points to the critical role these enzymes play in producing other biobased chemicals. It has been demonstrated that using POD from bitter gourd, a popular vegetable in China, makes a ferulic acid dehydrodimer with potent anti-inflammatory properties (Ou et al. 2003). Subsequent research on the enzyme's ability to oxidize ferulic acid also revealed the production of two additional dehydrodimers. Potent radical scavenging and cardioprotective effects were seen when a partially purified bitter gourd POD extract was applied to a ferulic acid/resveratrol combination. This led to synthesizing heterocoupling oligomers and dimers of ferulic acid and resveratrol (Yu et al. 2007).

6.2.2 Polyphenol Oxidases (PPOS)

PPOs, also called tyrosinases (EC 1.14.18.1), are oxidoreductases that contain copper and are found in many different organs within the same organism, including the leaves, fruits, and roots of higher plants. They are widely distributed throughout the phylogenetic scale, ranging from bacteria to mammals, and they even exhibit distinct biochemical characteristics in different organs (Yoruk and Marshall 2003). PPOs catalyze two reactions: the oxidation of the o-diphenol to create the equivalent o-quinone (diphenolase activity) and the hydroxylation of a monophenol to produce an o-diphenol (monophenolase activity). Molecular oxygen is needed for both processes. O-quinones are highly reactive molecules that can react with nucleophiles like amines and substances containing sulfhydryls or spontaneously polymerize into considerable molecular weight, insoluble brown solids without enzyme intervention. PPOs are a different class of enzymes that may be useful in several low-cost environmental operations.

It should be noted that various fungi and other microorganisms have been extensively researched for their potential to be used in bioremediation for several organic pollutants (Mukherjee et al. 2013). These organisms are highly effective at producing oxidative enzymes that utilize a broad range of potential substrates, including laccases (EC 1.10.3.1). However, multiple recent investigations have shown that PPOs recovered from plant food processing wastes are receiving more attention. Though a range of fruit and vegetable PPO-active homogenates have been tested for their ability to oxidize pollutants, such as bisphenol A, polyaromatic hydrocarbons (Lau et al. 2003), OMW phenolics (Greco Jr. et al. 1999), and indigo carmine, as well as producing bio-based substances like theaflavins (Tanaka et al. 2002). However, the spectrum of plant waste sources for PPO recovery is much more limited than that of PODs. As bioremediation agents, only PPOs derived from potatoes and mushrooms have undergone extensive testing; information on PPO derived from other sources is notably scarce. The availability and relatively wide range of substrates of mushroom PPO, which comes from processing residues (trims) of Agaricus bisporus and Pleurotus spp., have been the focus of previous studies. This enzyme treated common pollutants like phenol and chlorophenols in both model and industrial effluents when used in several forms (free, immobilized, crude extract, etc.). While a pH range of 5–7.8 was employed, in most cases, a strong removal efficacy was noted at pH 7.

6.3 Amylases

The two main classes of the amylase family are glucoamylase (G.A.) (EC 3.2.1.3) and amylase (EC 3.2.1.1). By cleaving the 1,4-a-D-glucosidic linkages between adjacent glucose units in the linear amylose chain, a-amylase can hydrolyze starch into maltose, glucose, and maltotriose (Pandey et al. 2000). In contrast, glucoamylase hydrolyzes the non-reducing ends of amylose and amylopectin to glucose. According to Pandey et al. (2000), amylases are extensively utilized in the food, fermentation, textile, and paper industries. They are also used to pretreat organic and agroindustrial by-products to increase the yield of bioproducts in the following operations. Through the optimization of fermentation employing various microbial strains, high-activity amylases may be generated from a variety of F.W.s, including kitchen waste (Wang et al. 2008), potato peel, coffee waste (Murthy et al. 2009), and tomato pomace (Umsza-Guez et al. 2011). However, because the activities of the generated enzymes are defined differently, it is challenging to compare the efficiency of different procedures. The primary benefit of producing enzymes from furrow wort (F.W.) is that it can be fermented without harsh pretreatments or additional nutritional supplements. Some studies showing the production of α-amylase from food waste are shown in Table 6.1.

According to Wang et al.'s (2008) investigation, glucoamylase synthesis from F.W. by Aspergillus niger UV-60 utilizing SmF was not significantly affected by nutritional supplementation. With 3.75% F.W. and 5% (v/w, 106 spores/mL) inoculum, the maximum glucoamylase activity of 137 U/mL was achieved at 30 C, 120 rpm for 96 h. With the generated crude glucoamylase, a lowering sugar concentration of 60.1 g/L could be achieved from 10% F.W. (w/v) in 125 min. Shukla and Kar (2006) also discovered that SSF produced high-activity amylase from potato peels. The alpha-amylase activity yielded by B. subtilis and B. licheniformis under ideal circumstances (40 C, pH 7, particle size of 1 mm, 65–70% moisture content) were 600 and 270 U/mL, respectively. A different work used a thermophilic isolate of alkaline-tolerant Bacillus firmus CAS7 strain to manufacture a-amylase from potato peels using SmF (Elayaraja et al. 2011). The ideal circumstances were 35 C, pH 7.5, and 1% substrate concentration. These resulted in 676 U/mL of a-amylase. Furthermore, Murthy et al. (2009) employed coffee wastes as the only carbon source to create a-amylase utilizing SSF and a Neurospora crassa CFR 308 fungal strain. The a-amylase activity of 4324 U/g dry substrate was measured at pH 4.6, 28 C, and 107 spores/g dry substrate with 60%

Table 6.1 Amylase production from food wastes

Residual materials	Microorganism	Fermentation conditions	Duration (day)	Achievements	References
Potato peel	Bacillus subtilis	40 C, pH 7, 65% MC, 10% (v/w) inoculum	2	α-amylase (600 U/mL)	Shukla and Kar (2006)
Coffee waste	Neurospora crassa CFR 308	28 C, pH 4.6, 60% MC, 1 mm P.S., 107 spores/g ds	5	α-amylase (6342 U/g ds)	Murthy et al. (2009)
Tomato pomace	Aspergillus awamori	28 C, pH 5	5	α-amylase (10.9 IU/g ds)	Umsza-Guez et al. (2011)
Pea pulp	Bacillus caldolyticus DSM 405	70 C, 150 rpm	6	α-amylase (8.6 U/mL)	Jamrath et al. (2012)
Bread waste	Aspergillus oryzae	30 C, MC:1.8 (w/w, db), PS:20 mm, 106 spore/gdS	6	GA (114 U/ gdS)	Melikoglu et al. (2013)
Wheat bran and potato peel substrate	Bacillus amyloliquefaciens	Phosphate buffer (10 mM phosphate, 262.0 µL of HAuCl4	2	α-amylase (1.3 U/ug)	Mojumdar and Deka (2019)

moisture content. Additionally, it was discovered that pretreating coffee waste with steam increased its accessibility and produced 6342 U/g of dry substrate with a-amylase activity.

6.4 Lignocellulolytic Enzymes

Many fungi primarily produce lignocellulolytic enzymes, including cellulases, xylanases, and ligninases. Degradation of the lignocellulosic materials is possible. Cellulases, for instance, are widely used in many different industries, such as agriculture, food, animal feed, brewing and wine production, biomass refining, pulp and paper, textiles, and washing (Kuhad et al. 2011). A complete cellulase system consisting of three enzyme classes must work in concert to bioconvert cellulose to fermentable sugars: endoglucanases (EC 3.2.1.4) that act randomly on soluble and insoluble cellulose chains; exoglucanases (cellobiohydrolases; EC 3.2.1.91) that release cellobiose from the reducing and non-reducing ends of cellulose chains; and b-glucosidases (EC 3.2.1.21) that release glucose from cellobiose (Jørgensen et al. 2007). With or without the concurrent usage of cellulases, xylanases have a wide range of uses in the food, feed, pulp and paper, brewing, wine-making, and textile sectors (Khandeparkar and Bhosle 2006). The primary enzymes needed for the hydrolysis of xylans are b-xylosidase and endo-b-1,4-xylanase. However, to hydrolyze substituted xylans, additional auxiliary enzymes must be present (Uc¸kun Kiran et al. 2013). Lignin is a polymer that is not wanted in biofuel synthesis because it makes plant-derived polysaccharides inaccessible. But lignin-derived materials can be used to create valuable products like adhesives, graft polymers like polyurethanes, polyesters, polyamines, epoxies, rubbers, dispersants, detergents, drilling mud thinner, surfactants, coagulants, flocculants (for treating sewage and wastewater) (Effendi et al. 2008). Ligninolytic enzyme combinations made up of laccases, lignin peroxidases, and Mn-peroxidase are used to break down lignin polymers. To increase the yields of bioproducts in later processes, lignocellulolytic enzymes are also utilized to pretreat agroindustrial and organic by-products (Menon and Rao 2012). Comparing the results and determining the most effective approach is challenging because there are various ways to define enzyme activity. But fungal SSF is generally the best approach (Bansal et al. 2012). According to Krishna (1999), SSF produced twelve times as much cellulase overall from banana waste as SmF. Díaz et al. (2012) discovered that SmF exhibited superior aeration compared to SSF, resulting in increased xylanase synthesis. Utilizing SSF in a plate-type bioreactor, Umsza-Guez et al. (2011) demonstrated a definite beneficial effect of aeration on the synthesis of xylanase and carboxymethyl cellulase (CMCase).

6.5 Pectinolytic Enzymes

Pectinases, or pectinolytic enzymes, break down pectin polymers through sequential and cooperative depolymerization and de-esterification processes. Endo- and exo-acting polygalacturonases, pectin- and pectate lyases, and rhamnogalacturonases—enzymes that cleave the rhamnogalacturonan chain—are necessary for the full breakdown of pectin (Kashyap et al. 2001). Pectinases are extensively utilized in the food sector, especially in the manufacture of juice and wine, as well as in several other industrial processes, including the textile, plant fiber processing, tea, coffee, oil extraction, and the treatment of industrial wastewater (Pedrolli et al. 2009). Aspergillus strains are beneficial for producing pectinases by fungal SSF (Kashyap et al. 2012). Due to the nature of these substrates and their low moisture content, pectinases can be produced industrially from pectin-containing wastes, such as citrus and orange wastes (Afifi 2011), apple pomace (Berovic and Ostroversnik 1997), grape pomace (Botella et al. 2005), or other fruit residues (Martínez et al. 2012), without the need for a harsh pretreatment. Hours et al.'s study from 1988 looked into the use of Aspergillus foetidus in SSF to produce pectinase from apple pomace. An enzyme activity of 1300 U/g was observed during a 36-h culture at 30 C with nitrogen supplementation. Aspergillus foetidus in SSF was also used to manufacture pectinolytic enzymes from citrus waste (Garzon and Hours 1992). After 36 h of growth, it was discovered that adding minerals and yeast extract significantly increased the activity. According to Berovic and Ostroversnik (1997), inexpensive nutrients, including soy flour, wheat bran, wheat maize, and whey, stimulated and enhanced the synthesis of pectolytic enzymes from apple pomace. Additionally, they stated that the most significant activity was attained at 38% initial moisture level and that moisture content was crucial for synthesizing enzymes. However, according to Ruiz et al. (2012), employing lemon peel pomace, 70% moisture content produced the maximum pectinase activity.

The viability of using grape pomace to help Aspergillus awamori produce exo-polygalacturonase in SSF was assessed by Botella et al. (2007). According to Hours et al. (1988), the particle size of the substrate did not affect the synthesis of enzymes; however, the addition of carbon sources and the initial moisture level of grape pomace was found to have a significant impact on the yields of enzymes. In a different work, Giese et al. (2008) used Botryosphaeria rhodina MAMB-05 in SSF and SmF with and without adding nutrients to produce pectinases from orange trash. In the SSF of orange bagasse, the maximum pectinase titer (32 U/mL) was obtained without adding fertilizers, and better microbial growth was seen. Another crucial factor influencing the synthesis of pectinase is aeration. According to Umsza-Guez et al. (2011), exo-PG synthesis was negatively impacted by forced aeration, which caused its activity in a multi-layer packed bed reactor to be cut in half. Maciel et al. (2013) obtained the greatest endo and exo-PG activity of 1.18 U/mL and 4.11 U/mL, respectively, using the reactor without aeration. An aeration-free system is better from an operational and financial standpoint. The medium's pH level may also impact the synthesis of pectinase. The influence of several substrates (apple, lemon peel, grape skin, and tamarind kernel) and fungi (Aspergillus flavipes FP-500 and Aspergillus terreus FP-370) on the generation of pectinases was studied by Martínez et al. (2012). With lemon peel, the maximum activity was attained. Higher pH values and lower carbon source concentration encouraged the development of pectin lyase and rhamnogalacturonase. In contrast, acidic pH values and high carbon source concentration supported the formation of exopeptidase and endopeptidase in both strains.

6.6 Proteases

Proteases are among the most significant commercial enzymes because of their numerous uses in the food, pharmaceutical, detergent, dairy, and leather sectors. According to reports, the most active protease producers are some strains of bacteria from the genus Bacillus and fungi Aspergillus, Penicillium, and Rhizopus. The use of F.W.s for protease production needs to be better documented, even though protease generation from agroindustrial wastes has been thoroughly investigated utilizing SSF and SmF. A recently identified alkalophilic Bacillus sp. was employed in SmF of date wastes by Khosravi-Darani et al. (2008). At pH 10, 37 C, high activity protease (57,420 APU/mL) was achieved. The enzyme was also found to be thermostable, suggesting it may be used in industrial applications. Afify et al. (2011) looked into using S. cerevisiae to produce proteases from potato waste in a submerged system and how to use the leftover solid waste as a biofertilizer to promote plant development. Using a fermentation medium containing 15 g of potato waste at an initial pH of 6.0, at 20 C for 72 h, the most significant enzyme activity (360 U/mg) was obtained. Using Bacillus pumilus MTCC 7514, fishmeal from sardine and pink perch was assessed as a sole carbon and nitrogen source in a study by Gupta et al. (2012). Compared to basal media (2646 U/mL), the protease achieved in a medium containing only fish meal (4914 U/mL) was almost twice as high. When the protease output was increased from a flask to 3.7 and 20 L fermenters, respectively, it reached 6966 and 7047 U/mL. When treating leather, it was discovered that the crude protease has dehairing abilities.

6.7 Lipases

Based on total sales volumes, lipases (EC 3.1.1.3) are regarded as the third largest group, behind proteases and carbohydrates (Contesini et al. 2010). They are widely utilized in the food, detergent, cosmetics, organic synthesis, and pharmaceutical industries for various purposes. Alkan et al. (2007) catalyze the hydrolysis of triacylglycerols to di- and mono-acylglycerols, fatty acids, and glycerol. Under specific circumstances, they can also catalyze the processes of alcoholysis, acidolysis, aminolysis, esterification, and transesterification. A subclass of phospholipases catalyzes the hydrolysis of one or more glycerophospholipid ester and phosphodiester linkages. Their modes of action on phospholipids differ, and they can be utilized to modify or create new phospholipids for usage in the food manufacturing, dairy, oil refinery, and cosmetics industries, among other sectors. Most studies have concentrated on generating high-activity extracellular lipase using SmF and SSF by various microorganisms, such as yeast, fungus, bacteria, and Actinomyces. Many strains of commercial lipase-producing fungi, such as those of Aspergillus, Yarrowia, Rhizopus, Rhizomucor, and Penicillium species, are highly prevalent (Colen et al. 2006). Several researchers have recently looked into the generation of lipase, either by adding F.W. as an inducer (Papanikolaou et al. 2011) or employing different F.W. as substrates (Alkan et al. 2007). Alkan et al.'s (2007) study explored using Bacillus coagulans in SSF to produce lipase from melon waste. After 24 h of cultivation with 1% olive oil enrichment at 37 C and pH 7, the maximum lipase production (78.1 U/g) was obtained by adding sodium dodecyl sulfate. Supplementing starch and maltose produced the most significant findings (148.9 and 141.6 U/g, respectively), while cultures grown on glucose and galactose showed relatively low enzyme activity (around 118.8 and 123.6 U/g, respectively). Mn^{2+} and Ni^{2+} inhibited enzymes by 68 and 74%, respectively.

6.8   Fish and Seafood Processing Wastes

These days, it is widely accepted that waste materials from aquatic species can be valued in several ways across several industries, including food products intended for human consumption. Up to 70% of fish and shellfish after industrial processing are considered by-products, and efforts to turn these into marketable goods have received much attention (Olsen et al. 2014). As per the Food and Agricultural Organization, approximately 27 million tonnes of marine biomass were utilized in 2008 for reasons other than food, such as bait, fish meal, fish oil, and the manufacturing of high-value chemicals for the cosmetic and pharmaceutical industries (Ordóñez-Del Pazo et al. 2014). Despite intensive research and development, Only a few highly value-added products based on fish processing wastes have made a lasting impression on the market. The availability of alternative, more affordable sources, the high costs of isolating specific components that are frequently present in small amounts in the by-products, the overestimation of market demand, and the scarcity of high-quality by-products that are regularly available are likely the leading causes of this (Olsen et al. 2014).

6.9   Conclusions

There are significant environmental and economic concerns about the management of F.W.s. This review emphasized how food waste can produce large quantities of industrial enzymes such as cellulose, lipases, proteases, amylases, and pectinases. The generated enzymes may be used in a few industrial applications depending on the purity requirements and F.W. transportation cost. Additionally, these enzymes can be included in other bioprocesses in biorefineries to create platform chemicals and biofuels. The conversion of F.W. into ethanol and other value-added products through biorefinery processes has only been accomplished at the bench-top and pilot levels thus far. Only a few F.W. biorefinery facilities are operating on an industrial scale. As a result, more than an economic analysis of the suggested biorefinery systems is needed. However, using inexpensive or free waste biomass could result in significant savings in operating expenses, given the cost of specified medium preparation in the existing commercial enzyme procedures. However, it's also essential to consider the challenges and costs of gathering and moving F.W. The larger-scale exploitation of F.W. for enzyme applications requires optimization and scale-up investigations.

References

Afifi MM (2011) Effective technological pectinase and cellulase by Saccharomyces cerevisiae utilizing food wastes for citric acid production. Life Sci J 8(2):405–413

Afify MM, Abd El-Ghany TM, Alawlaqi MM (2011) Microbial utilization of potato wastes for protease production and their using as biofertilizer. Aust J Basic Appl Sci 5(7):308–315

Al-Ansari MM, Steevensz A, Al-Aasm N, Taylor KE, Bewtra JK, Biswas N (2009) Soybean peroxidase catalyzed removal of phenylenediamines and benzenediols from water. Enzym Microb Technol 45:253–260

Alkan H, Baysal Z, Uyar F, Dogru M (2007) Production of lipase by a newly isolated Bacillus coagulans under solid-state fermentation using melon wastes. Appl Biochem Biotechnol 136(2):183–192

Bansal N, Tewari R, Soni R, Soni SK (2012) Production of cellulases from Aspergillus niger NS-2 in solid state fermentation on agricultural and kitchen waste residues. Waste Manag 32(7):1341–1346

Barakat N, Makris DP, Kefalas P, Psillakis E (2010) Removal of olive mill waste water phenolics using a crude peroxidase extract from onion by-products. Environ Chem Lett 8:271–275

Belcarz A, Ginalska G, Kowalewska B, Kulesza P (2008) Spring cabbage peroxidases – potential tool in biocatalysis and bioelectrocatalysis. Phytochemistry 69:627–636

Berovic M, Ostroversnik H (1997) Production of Aspergillus niger pectolytic enzymes by solid state bioprocessing of apple pomace. J Biotechnol 53(1):47–53

Botella C, de Orya I, Webbb C, Cantero D, Blandino A (2005) Hydrolytic enzyme production by Aspergillus awamori on grape pomace. Biochem Eng J 26(2–3):100–106

Botella C, Diaz A, de Ory I, Webb C, Blandino A (2007) Xylanase and pectinase production by Aspergillus awamori on grape pomace in solid state fermentation. Process Biochem 42(1):98–101

Colen G, Junqueira RG, Moraes-Santos T (2006) Isolation and screening of alkaline lipase-producing fungi from Brazilian savanna soil. World J Microbiol Biotechnol 22(8):881–885

Contesini FJ, Lopes DB, MacEdo GA, Nascimento MDG, Carvalho PDO (2010) Aspergillus sp. lipase: potential biocatalyst for industrial use. J Mol Catal B Enzym 67(3–4):163–171

Dec J, Bollag J-M (1994) Use of plant material for the decontamination of water polluted with phenols. Biotechnol Bioeng 44:1132–1139

Derat E, Shaik S (2006) An efficient proton-coupled electron-transfer process during oxidation of ferulic acid by horseradish peroxidase: coming complete cycle. J Am Chem Soc 128:13940–13949

Díaz AB, de Ory I, Caro I, Blandino A (2012) Enhance hydrolytic enzymes production by Aspergillus awamori on supplemented grape pomace. Food Bioprod Process 90(1):72–78

Duarte-Vázquez MA, Ortega-Tovar MA, García-Almendarez BE, Regalado C (2002) Removal of aqueous phenolic compounds from a model system by oxidative polymerization with turnip (Brassica napus L. var purple top white globe) peroxidase. J Chem Technol Biotechnol 78:42–47

Duarte-Vázquez MA, García-Padilla S, García-Almendarez BE, Whitaker JR, Regalado C (2007) Broccoli processing wastes as a source of peroxidase. J Agric Food Chem 55:10396–10404

Effendi A, Gerhauser H, Bridgwater AV (2008) Production of renewable phenolic resins by thermochemical conversion of biomass: a review. Renew Sust Energ Rev 12(8):2092–2116

Elayaraja S, Velvizhi T, Maharani V, Mayavu P, Vijayalakshmi S, Balasubramanian T (2011) Thermostable alphaamylase production by Bacillus firmus CAS 7 using potato peel as a substrate. Afr J Biotechnol 10(54):11235–11238

FAO wants to end hunger and transition to sustainable agricultural and food systems in the future. 2012, Food and Agriculture Organization of the United Nations, Rome

Gajalakshmi S, Abbasi SA (2008) Solid waste management by composting: state of the art. Crit Rev Environ Sci Technol 38(5):311–400

Garzon CG, Hours RA (1992) Citrus waste: an alternative substrate for pectinase production in solid-state culture. Bioresour Technol 39(1):93–95

Geng Z, Bassi AS, Gijzen M (2004) Enzymatic treatment of soils contaminated with phenol and chlorophenols using soybean seed hulls. Water Air Soil Pollut 154:151–166

Giese EC, Dekker RFH, Barbosa AM (2008) Orange bagasse as substrate for the production of pectinase and laccase by Botryosphaeria rhodina MAMB-05 in submerged and solid state fermentation. Bioresources 3(2):335–345

Greco G Jr, Toscano G, Cioffi M, Gianfreda L, Sannino F (1999) Dephenolisation of olive mill wastewaters by olive husk. Water Res 33:3046–3050

Gupta RK, Prasad D, Sathesh J, Naidu RB, Kamini NR, Palanivel S, Gowthaman MK (2012) Scale-up of an alkaline protease from Bacillus pumilus MTCC 7514 utilizing fish meal as a sole source of nutrients. J Microbiol Biotechnol 22(9):1230–1236

Hidalgo-Cuadrado N, Peérez-Galende P, Manzano T, De Maria CG, Shnyrov VL, Roig MG (2012) Screening of postharvest agricultural wastes as alternative sources of peroxidases: characterization and kinetics of a novel peroxidase from lentil (Lens culinaris L.) stubble. J Agric Food Chem 60:4765–4772

Hours RA, Voget CE, Ertola RJ (1988) Some factors affecting pectinase production from apple pomace in solid-state cultures. Biol Wastes 24(2):147–157

Jamrath T, Lindner C, Popovic MK, Bajpai R (2012) Production of amylases and proteases by Bacillus caldolyticus from food industry wastes. Food Technol Biotechnol 50(3):355–361

Jaramillo-Carmona S, Lopez S, Vazquez-Castilla S, Rodriguez-Arcos R, Jimenez-Araujo A, Guillen-Bejarano R (2013) Asparagus by-products as a new source of peroxidases. J Agric Food Chem 61:167–6174

Jørgensen H, Kristensen JB, Felby C (2007) Enzymatic conversion of lignocellulose into fermentable sugars: challenges and opportunities. Biofuels Bioprod Biorefin 1(2):119–134

Kashyap DR, Vohra PK, Chopra S, Tewari R (2001) Applications of pectinases in the commercial sector: a review. Bioresour Technol 77(3):215–227

Kashyap B, Das S, Kaur IR, Jhamb R, Jain S, Singal A, Gupta N (2012) Fungal profile of clinical specimens from a tertiary care hospital. Asian Pac J Trop Biomed 2(1):S401–S405

Khandeparkar RDS, Bhosle NB (2006) Isolation, purification and characterization of the xylanase produced by Arthrobacter sp. MTCC 5214 when grown in solid-state fermentation. Enzyme Microbial Technol 39(4):732–742

Khosravi-Darani K, Falahatpishe HR, Jalali M (2008) Alkaline protease production on date waste by an alkalophilic Bacillus sp. 2-5 isolated from soil. Afr J Biotechnol 7(10):1536–1542

Kiran E, Akpinar O, Bakir U (2013) Improvement of enzymatic xylooligosaccharides production by the co-utilization of xylans from different origins. Food Bioprod Process 91(4):565–574

Klibanov AM, Morris ED (1981) Horseradish peroxidase for the removal of carcinogenic aromatic amines from water. Enzym Microb Technol 3:119–122

Krishna C (1999) Production of bacterial cellulases by solid state bioprocessing of banana wastes. Bioresour Technol 69(3):231–239

Kuhad RC, Gupta R, Singh A (2011) Microbial cellulases and their industrial applications. Enzyme Res 2011(1):10

Lau KL, Tsang YY, Chiu SW (2003) Use of spent mushroom compost to bioremediate PAH-contaminated samples. Chemosphere 52:1539–1546

Lin CSK, Pfaltzgraff LA, Herrero-Davila L, Mubofu EB, Abderrahim S, Clark JH, Koutinas AA, Kopsahelis N, Stamatelatou K, Dickson F, Thankappan S, Mohamed Z, Brocklesby R, Luque R (2013) Food waste as a valuable resource for the production of chemicals, materials and fuels. Current situation and global perspective. Energy Environ Sci 6(2):426–464

López-Molina D, Hiner ANP, Tudela J, García-Cánovas F, Rodríguez-López HN (2003) Enzymatic removal of phenols from aqueous solution by artichoke (Cynara scolymus L.) extracts. Enzym Microb Technol 33:738–742

Lundqvist J, de Fraiture C, Molden D (2008) Saving water: from field to fork: curbing losses and wastage in the food chain, in SIWI policy brief. Stockholm International Water Institute Stockholm, Sweden

Maciel M, Ottoni C, Santos C, Lima N, Moreira K, Souza-Motta C (2013) Production of polygalacturonases by Aspergillus section Nigri strains in a fixed bed reactor. Molecules 18(2):1660–1671

Martínez M, Yáñez R, Alonsó JL, Parajó JC (2012) Pectic oligosaccharides production from orange peel waste by enzymatic hydrolysis. Int J Food Sci Technol 47(4):747–754

Melikoglu M, Lin CSK, Webb C (2013) Stepwise optimisation of enzyme production in solid state fermentation of waste bread pieces. Food Bioprod Process 91(4):638–646

Menon V, Rao M (2012) Trends in bioconversion of lignocellulose: biofuels, platform chemicals & biorefinery concept. Prog Energy Combust Sci 38(4):522–550

Mojumdar A, Deka J (2019) Recycling agro-industrial waste to produce amylase and characterizing amylase–gold nanoparticle composite. Int J Recycl Org Waste Agric 8. https://doi.org/10.1007/s40093-019-00298-4

Mukherjee S, Basak B, Bhunia B, Dey A, Mondal B (2013) Potential use of polyphenol oxidases (PPO) in the bioremediation of phenolic contaminants containing industrial wastewater. Rev Environ Sci Biotechnol 12:61–73

Murthy PS, Madhava Naidu M, Srinivas P (2009) Production of aamylase under solid-state fermentation utilizing coffee waste. J Chem Technol Biotechnol 84(8):1246–1249

Ngoc UN, Schnitzer H (2009) Sustainable solutions for solid waste management in Southeast Asian countries. Waste Manag 29:1982–1995

Olsen RL, Toppe J, Karunasagar I (2014) Challenges and realistic opportunities in the use of by-products from processing of fish and shellfish. Trends Food Sci Technol 36:144–151

Ordóñez-Del Pazo T, Antelo LT, Franco-Uría A, Pérez-Martín RI, Sotelo CG, Alonso AA (2014) Fish discards management in selected Spanish and Portuguese métiers: identification and potential valorisation. Trends Food Sci Technol 36:29–43

Othman SN, Noor ZZ, Abba AH, Yusuf RO, Abu MA (2013) Hassan, review on life cycle assessment of integrated solid waste management in some Asian countries. J Clean Prod 41: 251–262

Ou L, Kong L-Y, Zhang X-M, Niwa M (2003) Oxidation of ferulic acid by Momordica charantia peroxidase and related anti-inflammation activity changes. Biol Pharm Bull 26:1511–1516

Paice MG, Jurasek L (1984) Peroxidase-catalyzed color removal from bleach plant effluent. Biotechnol Bioeng 26:477–480

Pandey A, Nigam P, Soccol CR, Soccol VT, Singh D, Mohan R (2000) Advances in microbial amylases. Biotechnol Appl Biochem 31(2):135–152

Papanikolaou S, Dimou A, Fakas S, Diamantopoulou P, Philippoussis A, Galiotou-Panayotou M, Aggelis G (2011) Biotechnological conversion of waste cooking olive oil into lipidrich biomass using Aspergillus and Penicillium strains. J Appl Microbiol 110(5):1138–1150

Pedrolli DB, Monteiro AC, Gomes E, Carmona EC (2009) Pectin and pectinases: production, characterization and industrial application of microbial pectinolytic enzymes. Open Biotechnol J 3:9–18

Ruiz HA, Rodriguez-Jasso RM, Rodriguez R, ContrerasEsquivel JC, Aguilar CN (2012) Pectinase production from lemon peel pomace as support and carbon source in solid state fermentation column-tray bioreactor. Biochem Eng J 65:90–95

Satar R, Husain Q (2009) Use of bitter gourd (Momordica charantia) peroxidase together with redox mediators to decolorize disperse dyes. Biotechnol Bioprocess Eng 14:213–219

Shukla J, Kar R (2006) Potato peel as a solid state substrate for thermostable alpha amylase production by thermophilic Bacillus isolates. World J Microbiol Biotechnol 22(5):417–422

Takata M, Fukushima K, Kino-Kimata N, Nagao N, Niwa C, Toda T (2012) The effects of recycling loops in food waste management in Japan: based on the environmental and economic evaluation of food recycling. Sci Total Environ 432:309–317

Tanaka T, Mine C, Inoue K, Matsuda M, Kouno I (2002) Synthesis of theaflavin from epicatechin and epigallocatechin by plant homogenates and role of epicatechin quinone in the synthesis and degradation of theaflavins. J Agric Food Chem 50:2142–2148

Umsza-Guez MA, Díaz AB, de Ory I, Blandino A, Gomes E, Caro I (2011) Xylanase production by Aspergillus awamori under solid state fermentation conditions on tomato pomace. Braz J Microbiol 42(4):1585–1597

Wang Q, Wang X, Wang X, Ma H (2008) Glucoamylase production from food waste by Aspergillus niger under submerged fermentation. Process Biochem 43(3):280–286

Yoruk R, Marshall MR (2003) Physicochemical properties and function of plant polyphenol oxidase: a review. J Food Biochem 27:361–422

Yu B-B, Han X-Z, Lou H-X (2007) Oligomers of resveratrol and ferulic acid prepared by peroxidase catalyzed oxidation and their protective effects on cardiac injury. J Agric Food Chem 55:7753–7757